Organolithium Methods

BEST SYNTHETIC METHODS

Series Editors

A. R. Katritzky
University of Florida
Gainesville, Florida
USA

O. Meth-Cohn
Sterling Organics Ltd
Newcastle-upon-Tyne
UK

C. W. Rees
Imperial College of Science
and Technology
London, UK

Richard F. Heck, *Palladium Reagents in Organic Syntheses*, 1985
Alan H. Haines, *Methods for the Oxidation of Organic Compounds: Alkanes, Alkenes, Alkynes, and Arenes*, 1985
Paul N. Rylander, *Hydrogenation Methods*, 1985
Ernest W. Colvin, *Silicon Reagents in Organic Synthesis*, 1988
Andrew Pelter, Keith Smith and Herbert C. Brown, *Borane Reagents*, 1988
Basil Wakefield, *Organolithium Methods*, 1988

In preparation

Alan H. Haines, *Methods for the Oxidation of Organic Compounds: Alcohols, Alcohol Derivatives, Alkyl Halides, Nitro Alkenes, Alkyl Azides, Carbonyl Compounds, Hydroxyarenes, and Amino Arenes*, 1988.
I. Ninomiya and T. Naito, *Photochemical Synthesis*, 1988

Organolithium Methods

Basil J. Wakefield
Department of Chemistry and Applied Chemistry
University of Salford, Salford M5 4WT
UK

1988

Academic Press
Harcourt Brace Jovanovich, Publishers
London San Diego New York Boston
Sydney Tokyo Toronto

ACADEMIC PRESS LIMITED
24–28 Oval Road
London NW1 7DX

US Edition published by
ACADEMIC PRESS INC.
San Diego, CA 92101

Copyright © 1988 by
ACADEMIC PRESS LIMITED

All rights Reserved

No part of this book may be reproduced in any form by photostat, microfilm, or any other means, without written permission from the publishers

This book is a guide providing general information concerning its subject matter; it is not a procedural manual. Synthesis of chemicals is a rapidly changing field. The reader should consult current procedural manuals for state-of-the-art instructions and applicable government safety regulations. The publisher and the authors do not accept responsibility for any misuse of this book, including its use as a procedural manual or as a source of specific instructions

British Library Cataloguing in Publication Data

Wakefield, B. J.
 Organolithium methods.
 1. Organic compounds. Synthesis. Use of organic lithium compounds
 I. Title   II. Series
 547.2

ISBN 0-12-730940-3

Typeset by Blackpool Typesetting Services Limited, Blackpool
Printed in Great Britain by St Edmundsbury Press Limited, Bury St Edmunds, Suffolk

Contents

Foreword vii

Preface ix

Detailed Contents xi

Abbreviations xvii

Chapter 1.	Introduction	1
Chapter 2.	General Considerations in the Application of Organolithium Compounds in Organic and Organometallic Synthesis	3
Chapter 3.	Preparation of Organolithium Compounds	21
Chapter 4.	Addition of Organolithium Compounds to Carbon–Carbon Multiple Bonds	53
Chapter 5.	Addition of Organolithium Compounds to Carbon–Nitrogen Multiple Bonds	57
Chapter 6.	Addition of Organolithium Compounds to Carbonyl Groups	67
Chapter 7.	Addition of Organolithium Compounds to Thiocarbonyl Groups	101
Chapter 8.	Substitution at Carbon by Organolithium Compounds . . .	107
Chapter 9.	Reactions of Organolithium Compounds with Proton Donors	119
Chapter 10.	Formation of Carbon–Nitrogen Bonds via Organolithium Compounds	125
Chapter 11.	Formation of Carbon–Oxygen Bonds via Organolithium Compounds	129
Chapter 12.	Formation of Carbon–Sulphur Bonds via Organolithium Compounds	135
Chapter 13.	Formation of Carbon–Halogen Bonds via Organolithium Compounds	143
Chapter 14.	Synthesis of Organoboron, Organosilicon and Organophosphorus Compounds from Organolithium Compounds	149
Chapter 15.	Organolithium Compounds in the Synthesis of other Organometallic Compounds	159

Chapter 16. Applications of Elimination Reactions of Organolithium Compounds; Arynes, Carbenes, Ylides, Ring Opening of Heterocycles 167

Index of Compounds and Methods 181

Foreword

There is a vast and often bewildering array of synthetic methods and reagents available to organic chemists today. Many chemists have their own favoured methods, old and new, for standard transformations, and these can vary considerably from one laboratory to another. New and unfamiliar methods may well allow a particular synthetic step to be done more readily and in higher yield, but there is always some energy barrier associated with their use for the first time. Furthermore, the very wealth of possibilities creates an information-retrieval problem. How can we choose between all the alternatives, and what are their real advantages and limitations? Where can we find the precise experimental details, so often taken for granted by the experts? There is therefore a constant demand for books on synthetic methods, especially the more practical ones like *Organic Syntheses*, *Organic Reactions*, and *Reagents for Organic Synthesis*, which are found in most chemistry laboratories. We are convinced that there is a further need, still largely unfulfilled, for a uniform series of books, each dealing concisely with a particular topic from a *practical* point of view—a need, that is, for books full of preparations, practical hints and detailed examples, all critically assessed, and giving just the information needed to smooth our way painlessly into the unfamiliar territory. Such books would obviously be a great help to research students as well as to established organic chemists.

We have been very fortunate with the highly experienced and expert organic chemists who, agreeing with our objective, have written the first group of volumes in this series, *Best Synthetic Methods*. We shall always be pleased to receive comments from readers and suggestions for future volumes.

A.R.K., O.M.-C., C.W.R.

Preface

My aim in compiling this book has been to include only reliable, well-described procedures, whether as detailed examples or as entries in the tables. In my task I have been helped by colleagues from other institutions, who have generously provided me with excellent experimental details. I also thank colleagues at Salford and members of my research group, who have confirmed the reliability of many of the procedures from their own experience. All chemists are of course greatly indebted to the submitters, checkers and editors of *Organic Syntheses* and *Inorganic Syntheses* for publishing checked procedures.

In cases where the expertise of others has not been available to me I have relied on my own judgement of published procedures. If my judgement has been at fault, I shall be grateful for any relevant information.

B. J. Wakefield
Salford

Detailed Contents

1. **Introduction** 1

2. **General Considerations in the Application of Organolithium Compounds in Organic and Organometallic Synthesis**
   - 2.1. Constitution and Reactivity of Organolithium Compounds 3
   - 2.2. Practical Considerations 4
     - 2.2.1. Choice of Solvent 4
     - 2.2.2. Reactions at Low Temperatures 8
     - 2.2.3. Inert Atmospheres 9
     - 2.2.4. Storage and Transfer of Organolithium Compounds; Safety Precautions 11
     - 2.2.5. Typical Assemblies for Reactions 16
   - 2.3. Detection and Estimation of Organolithium Compounds 16
     - 2.3.1. Detection 16
       - *Gilman Colour Tests* 16
     - 2.3.2. Estimation 17
       - *Estimation of n-butyllithium in hydrocarbons by Gilman Double Titration* 18
   - References 19

3. **Preparation of Organolithium Compounds**
   - 3.1. Preparation from Organic Halides 21
     - 3.1.1. By Reaction with Lithium Metal 21
       - *t-Butyllithium* 23
       - *Phenyllithium* 24
       - *Ethyllithium* 24
     - References 25
     - 3.1.2. By Reaction with Lithium Salts of Radical Anions 26
       - *Lithium 4,4'-di-t-butylbiphenyl (LiDBB)* 26
       - *7-Lithionorbornadiene* 27
     - References 27
     - 3.1.3. By Metal–Halogen Exchange 27
       - *1-Hexenyllithium* 29
       - *Pentachlorophenyllithium* 29
       - *3-Lithiobenzo[b]thiophene* 31
     - References 31

3.2. Preparation by Metallation 32
        2-Lithio-1,3-dithiane 36
        A mixed aldol reaction with cyclopentanone enolate 37
        Benzyllithium 38
        2-Lithiothiophene and 2,5-dilithiothiophene 38
        2-Lithio-N,N-diethylbenzamide 39
        Metallation of propene by butyllithium–potassium t-butoxide reagent 39
        References 43
3.3. Preparation from other Organometallic Compounds 44
        1,2-Dilithiobenzene 45
        Vinyllithium 46
        References 47
3.4. Preparation from Ethers and Thioethers 47
        Allyllithium 47
        4-t-Butyl-1-phenylthiocyclohexyllithium 48
        References 49
3.5. Preparation from Sulphonylhydrazones 49
        2-Lithiobornene 50
        References 50
3.6. Preparation by Addition to Carbon–Carbon and Carbon–Sulphur Multiple Bonds 51

4. **Addition of Organolithium Compounds to Carbon–Carbon Multiple Bonds**
        2-(2,2-Dimethylpropyl)-2-lithio-1,3-dithiane by addition of t-butyllithium to 2-methylene-1,3-dithiane 54
        Addition of n-butyllithium to allyl alcohol; 2-methylhexan-1-ol .. 54
        Reaction of lithiated O-(1-ethoxyethyl) 2-methylpropanal cyanohydrin with (η^6-2-methylanisole)tricarbonylchromium; 2-methyl-5-(2-methylpropanoyl)anisole 55
        References 56

5. **Addition of Organolithium Compounds to Carbon–Nitrogen Multiple Bonds**
   5.1. Addition to Imines and Iminium Salts 57
        Addition of n-butyllithium to dimethyl(methylene)ammonium iodide (Eschenmoser's salt) 58
        References 58
   5.2. Addition to Nitrogen Heterocyclic Aromatic Compounds 58
        Reactions proceeding via addition of phenyllithium to pyridine .. 60
        References 61
   5.3. Addition to Nitriles 62
        Dicyclopentyl ketone 64
        Di-t-butylketimine 64
        References 65
   5.4. Addition to Isonitriles 65
        4-Lithio-3,6,6,8,8-pentamethyl-5-azanon-4-ene 66
        References 66

DETAILED CONTENTS xiii

6. **Addition of Organolithium Compounds to Carbonyl Groups**

6.1. Addition to Aldehydes and Ketones 67
    6.1.1. Addition to Saturated and Aryl Aldehydes and Ketones 67
            Addition of allyllithium to 4-methylpentan-2-one 70
            3-t-Butyl-3-hydroxy-2,2-dimethylheneicosane by a Barbier-type reaction 70
            References 71
    6.1.2. Addition to $\alpha\beta$-Unsaturated Aldehydes and Ketones 71
            Conjugate addition of α-lithio(4-methoxyphenyl)acetonitrile to cyclohex-2-enone 72
            Addition of a cyanocuprate reagent; 4-methyl-4-phenylpentan-2-one 73
            Addition of lithium dibutylcuprate to cyclohexenone, followed by reaction with iodomethane; 3-butyl-2-methylcyclohexanone 73
            References 74
    6.1.3. Peterson Olefination and Related Reactions 74
            Diphenyl(2,2-diphenylethenyl)phosphine sulphide 75
            References 75
6.2. Reactions with Acyl Halides, Anhydrides, Esters and Lactones 76
        Reaction with an anhydride; 2,2-dichloro-4-methylpentan-3-one . 80
        Reaction with methyl formate; 2,2-dichloro-2-phenylethanal . . . 81
        References 82
6.3. Addition to *N,N*-Disubstituted Amides 82
        Reaction with DMF: 3-chloro-2-(methoxymethoxy)benzaldehyde . 87
        References 87
6.4. Addition to Cumulated Carbonyl Groups and Carboxylate Ions 88
        Reaction with carbon dioxide; 2-chloro-3,3-diphenylacrylic acid . 92
        Reaction with a carboxylate; acetylcyclohexane 93
        References 94
6.5. Addition to Carbon Monoxide and Metal Carbonyls 95
    6.5.1. Addition to Carbon Monoxide; Generation and Trapping of Acyllithium Compounds 95
            Generation and trapping of an acyllithium compound; 3-hydroxy-2,2,3-trimethyloctan-4-one 97
            References 97
    6.5.2. Addition to Metal Carbonyls 98
            Addition to chromium hexacarbonyl; pentacarbonyl [methoxy(1,4-dimethoxy-2-naphthyl)carbene] chromium(0) 99
            References 99

7. **Addition of Organolithium Compounds to Thiocarbonyl Groups**

7.1. Carbophilic Addition 101
        Reaction with carbon disulphide; methyl 2-carboxydithiohexadecanoate 103
        References 104
7.2. Thiophilic Addition 104
        References 105

8. Substitution at Carbon by Organolithium Compounds

- 8.1. Displacement of Halide 107
  - *Synthesis of 3-butylpyridine from 3-(lithiomethyl)pyridine and 1-bromopropane* 108
  - *Syntheses via lithiated 1,3-dithianes; 5,9-dithiaspiro[3.5]nonane* 110
  - References 110
- 8.2. Displacement of Sulphate and Sulphonate 111
  - *Alkylation of an arenesulphonate; 2-ethyl-1,3-dithiane* 112
  - References 112
- 8.3. Nucleophilic Ring Opening of Epoxides and Other Cyclic Ethers 113
  - *Reaction with oxirane; (E)-4-ethyloct-3-en-1-ol* 113
  - *Ring opening of an epoxide promoted by boron trifluoride; trans-2-butylcyclohexanol* 116
  - *Reaction with oxetane; 3-pentachlorophenylpropan-1-ol* ... 117
  - References 117

9. Reactions of Organolithium Compounds with Proton Donors

- 9.1. Formation of Lithium Alkoxides, Thiolates and Amides 119
  - References 119
- 9.2. Hydrogen Isotopic Labelling via Organolithium Compounds 120
  - *N-(1-d-2-Methylbutylidene)-1,1,3,3-tetramethylbutylamine and 1-d-2-methylbutanal* 121
  - References 123
- 9.3. Indirect Reduction of Organic Halides 123
  - Reference 124

10. Formation of Carbon–Nitrogen Bonds via Organolithium Compounds

- 10.1. Reactions with Hydroxylamine Derivatives 125
  - *Amination of phenyllithium* 126
  - References 126
- 10.2. Reactions with Azides 127
  - *Reaction with a sulphonyl azide; 2-azido-3,3'-bithienyl and 2-amino-3,3'-bithienyl* 127
  - References 128

11. Formation of Carbon–Oxygen Bonds via Organolithium Compounds

- 11.1. Reaction with Dioxygen 129
  - *Reaction with dioxygen; α-hydroxyphenylacetic acid and α-hydroperoxyphenylacetic acid* 130
  - References 131
- 11.2. Reaction with Peroxides 131
  - References 132
- 11.3. Reaction with Molybdenum Pentoxide–Pyridine–Hexametapol 133
  - References 133

12. Formation of Carbon–Sulphur Bonds via Organolithium Compounds

- 12.1. Reaction with Elemental Sulphur 135
  - *Thiophene-2-thiol* 136
  - References 137

DETAILED CONTENTS XV

12.2. Reaction with Disulphides 137
E-*1-Phenylthio-1-hexene* 137
References 139
12.3. Reaction with Sulphur Halides and Related Compounds 139
References 140
12.4. Reaction with Compounds Containing Sulphur–Oxygen Double Bonds 141
Reaction with sulphur dioxide; lithium n-butylsulphinate ... 142
References 142

13. **Formation of Carbon–Halogen Bonds via Organolithium Compounds**

Reaction with hexachloroethane; 3-chlorofuran 147
Reaction with elemental iodine; pentachloroiodobenzene ... 147
References 147

14. **Synthesis of Organoboron, Organosilicon and Organophosphorus Compounds from Organolithium Compounds**

14.1. Synthesis and Oxidation of Organoboron Compounds 149
Synthesis of benzo[b]*thiophen-2(3H)-one by formation and oxidation of an organoboron compound* 151
References 152
14.2. Synthesis of Organosilicon Compounds 152
Reaction with chlorotrimethylsilane; 1-phenylthio-1-trimethylsilylethene 154
References 154
14.3. Synthesis of Organophosphorus Compounds 155
14.3.1. Phosphorus(III) Compounds 155
Reaction with a halophosphine; [2-(methylthio)phenyl]diphenylphosphine 155
14.3.2. Phosphorus(v) Compounds 157
References 157

15. **Organolithium Compounds in the Synthesis of other Organometallic Compounds**

Methyldiphenylbismuth 163
cis-*Bis[4-(trimethylsilyl)phenyl]bis(triphenylphosphine)platinum* 164
References 164

16. **Applications of Elimination Reactions of Organolithium Compounds; Arynes, Carbenes, Ylides, Ring Opening of Heterocycles**

16.1. Organolithium Compounds as Precursors of Arynes 167
Generation and trapping of tetrachlorobenzyne; 5,8-epoxy-1,2,3,4-tetrachloro-5,8-dihydronaphthalene 168
References 169
16.2. Organolithium Carbenoid Reactions 169
16.2.1. Synthesis of Cyclopropanes and Oxiranes 169
Carbenoid cycloaddition via dibromofluoromethyllithium; 1-bromo-1-fluoro-2,2,3,3-tetramethylcyclopropane 170

16.2.2. Organolithium Carbenoid Insertion Reactions 172
 1,6-Dimethyltricyclo[4.1.0.0^{2,7}]hept-3-ene 172
16.2.3. Organolithium Carbenoid Rearrangement Reactions 173
 References 174
16.3. Organolithium Compounds in the Generation of Ylides 174
 Selective formation of an E-alkene by a modified Wittig
 reaction; E-oct-2-ene 175
 References 176
16.4. Syntheses via Ring Opening of Lithiated Heterocycles 176
 Synthesis of a selenoenyne by ring opening of
 lithioselenophene 179
 References 179

Abbreviations

DABCO	1,4-diazabicyclo[2.2.2]octane
DME	1,2-dimethoxyethane
DMF	N,N-dimethylformamide
DMSO	dimethylsulphoxide
ether	diethyl ether
hexametapol*	hexamethylphosphoric triamide, tris(dimethylamino)phosphine oxide
LDA	lithium diisopropylamide
LDMAN	lithium 1-dimethylaminonaphthalene
LiDBB	lithium 4,4'-di-t-butylbiphenyl
LiTMP	lithium 2,2,6,6-tetramethylpiperidide
LN	lithium naphthalene
MoOPH	molybdenum pentoxide–pyridine–hexametapol complex
THF	tetrahydrofuran
TMEDA	N,N,N',N'-tetramethyl-1,2-ethanediamine, tetramethylethylenediamine
°	°C

*Confusion has been caused by the inconsistent use of the abbreviations HMPT/HMPA for $(Me_2N)_3P/(Me_2N)_3PO$. The use of Normant's trivial name "hexametapol" for the phosphorus(v) compound avoids ambiguity.

—1—
Introduction

Organolithium compounds are such versatile and widely used reagents for organic synthesis that only a very limited selection of reactions and procedures can be described within a modestly sized book. Nevertheless, the examples have been chosen to be representative of the most important types of reaction. These include not only the fundamental carbon–carbon bond-forming reactions, but also reactions for introducing a wide range of functional groups. For further information on the chemistry of organolithium compounds, including discussions of their structures and constitution and of the mechanisms of their reactions, the following major sources may be consulted. These works are referred to in later chapters as General Refs A–D.

A B. J. Wakefield, *The Chemistry of Organolithium Compounds*, Pergamon, Oxford, 1974.
B *Methoden der Organischen Chemie* (Houben–Weyl), Vol. XIII/1, 4th edn, Thieme, Stuttgart, 1970.
C M. Schlosser, *Polare Organometalle*, Springer, Berlin, 1973.
D (i) J. L. Wardell, Chapter 2, and (iii) B. J. Wakefield, Chapter 44, in *Comprehensive Organometallic Chemistry*, ed. G. Wilkinson, Pergamon, Oxford, 1982.

Basic information on the more important individual organolithium compounds is listed, with leading references, in General Ref.

E *Dictionary of Organometallic Compounds*, ed. J. Buckingham, Chapman and Hall, London, 1984.

The chemistry of organolithium compounds is still developing rapidly, and new applications are continually being reported. Recent developments are reviewed annually in the Annual Surveys of the *Journal of Organometallic Chemistry* and, more briefly, in the Specialist Periodical Report, *Organometallic Chemistry* of the Royal Society of Chemistry. The *Dictionary of Organometallic Compounds* (general reference E) is also supplemented annually.

—2—
General Considerations in the Application of Organolithium Compounds in Organic and Organometallic Synthesis

2.1. CONSTITUTION AND REACTIVITY OF ORGANOLITHIUM COMPOUNDS

Considering their reactivity (and in some cases their thermal instability), a surprising number of organolithium compounds have been obtained as pure liquids or crystalline solids. Examples of their physical properties are listed in General Refs A and D(i), and a more extensive listing is given in General Ref. E and supplements. Recent work on the often remarkable structural features of organolithium compounds has been reviewed (1-3). However, when they are employed in synthesis, organolithium compounds are almost invariably prepared and used in solution.

Many organolithium compounds are soluble in hydrocarbons; important exceptions are methyllithium and phenyllithium. In hydrocarbon solution they are associated; for example, n-butyllithium in cyclohexane is largely hexameric, and t-butyllithium is tetrameric in the same solvent. The bonding in the associated species is electron-deficient, so that in their reactions they can behave, paradoxically, as both electron-poor reagents (Lewis acids) and as carbanionoid reagents (Brønsted bases, nucleophiles).

A Lewis base can interact with an organolithium oligomer simply by coordination with the electron-deficient framework, but very often it causes a decrease in the degree of association. In general, the stronger the electron donor, the lower the resulting degree of association, and with very strong donors, particularly difunctional donors, even monomeric complexes may result. Thus methyllithium, which is tetrameric in the solid phase, becomes a solvated tetramer in diethyl ether; and n-butyllithium, hexameric in hydrocarbons, becomes tetrameric in diethyl ether. In the more strongly basic THF, recent low-temperature studies have shown n-butyllithium to have a degree of association between dimeric and trimeric at $-108°$, and phenyllithium to be between monomeric and dimeric (4). The complexes of

3

organolithium compounds with strong, difunctional electron donors such as TMEDA and DABCO are often comparatively insoluble, so, although the structures of several have been determined by X-ray crystallography, little is known about their degree of association in solution.

The effect of the solvent and/or the presence of electron donors may be summarized, in an admittedly over-simplified form, as follows. In a hydrocarbon solvent, and in the absence of electron-donating ligands, the electron deficiency of the reagent is maximized. This factor may, for example, be important in determining the stereochemistry of the polymerization of dienes, initiated by organolithium compounds (see General Ref. A for a brief account).

The effect of electron donor solvents or additives is to reduce the electron deficiency of the reagents, while at the same time increasing their carbanionoid character, and thus both their nucleophilicity and their basicity. The latter effect is particularly manifested in the promotion of metallation by additives such as TMEDA (see Section 3.2). In the extreme case, ion pairs of carbanions and solvated lithium ions are probably present in solutions, though in the solid state there is a close association between lithium and the "carbanion" even when the latter is delocalized.

There is growing evidence for a single-electron-transfer component in some "nucleophilic" reactions of organolithium compounds. Such pathways are also promoted by cation-solvating solvents or additives, both by increasing the carbanionic character of the reagents and by solvating electrons. For example, the promotion of conjugate addition of organolithium compounds to $\alpha\beta$-unsaturated carbonyl compounds by the presence of hexametapol (see p. 72) may be attributed to such effects.

2.2. PRACTICAL CONSIDERATIONS

2.2.1. Choice of Solvent

Besides the "chemical" factors outlined above, some "practical" factors have to be taken into account in choosing the solvent for an organolithium reagent. The solvent should be easily purified and freed from water and peroxides. It should have an appropriate boiling point or—usually more important—a low-enough freezing point. And it should not itself react with organolithium compounds.

The physical properties of the most useful solvents for organolithium compounds are listed in Table 2.1, and comments on the individual solvents follow.

2.2. PRACTICAL CONSIDERATIONS

TABLE 2.1
Solvents for Organolithium Reactions

Solvent	b.p.	m.p.
Pentane	36°	−129°
Hexane	69°	−94°
Benzene	80°	5.5°
Diethyl ether	34.5°	−116°
THF	64–5°	−65°
DME	82–3°	−58°
Triethylamine	89°	−115°
TMEDA	122°	−55°
Hexametapol	232°	7°

Aliphatic hydrocarbons are inert, and easily purified and dried. Besides those listed, others and mixed "light petroleum" fractions may be used. The isomeric butyllithiums are commercially available in these solvents. However, aryllithium compounds, methyllithium, and reagents such as LDA are insoluble in aliphatic hydrocarbons in the absence of electron-donating ligands.

Aromatic hydrocarbons are also easily purified and dried, but toluene is relatively easily lithiated, and benzene is dangerously toxic and cannot be used at low temperatures. They may, however, be useful in particular cases where the presence of ethers etc. is undesirable and the solubility of the organolithium compound in aliphatic hydrocarbons is too low. For example, ethyllithium is considerably more soluble in benzene than in pentane.

Ethers are the most commonly used solvents for organolithium reactions. Diethyl ether is a good general-purpose choice. Most organolithium compounds are soluble in it, and it is not in most cases too rapidly cleaved by them; it may be used at low temperatures; and it is a convenient solvent for conventional work-up procedures. In many cases it is adequately dried by sodium wire, and its high vapour pressure at room temperature creates a "blanket" over the surface of the solution, which sometimes makes provision of a nitrogen or argon atmosphere unnecessary. Nevertheless, for low concentrations of organolithium compounds or for work at low temperatures, drying and handling procedures like those described below must be used. The hazardous properties of diethyl ether—its flammability, proneness to the formation of peroxides, and toxicity—are well known, and appropriate safety precautions should be taken.

When a more strongly Lewis-basic solvent is required, THF is the most widely used alternative to diethyl ether. It does, however, suffer from some disadvantages. It is much more readily attacked by organolithium

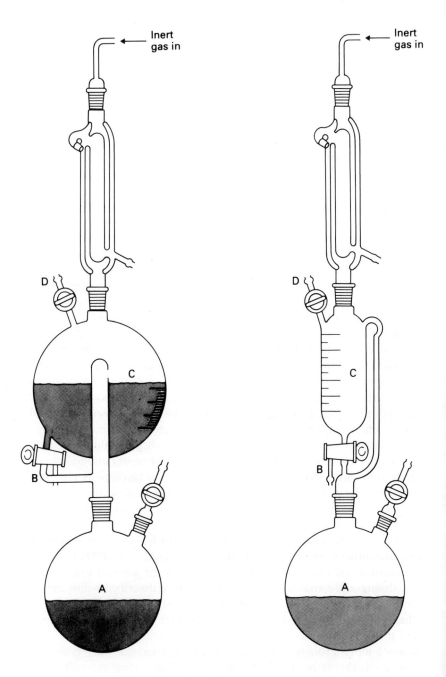

Fig. 2.1. Stills for preparing dry THF and other solvents.

2.2. PRACTICAL CONSIDERATIONS

compounds than diethyl ether; its freezing point is only just above the temperature of evaporating solid carbon dioxide; and it is much less easily dried, and more hygroscopic, than diethyl ether. Drying procedures that have been found satisfactory for THF include distillation from lithium aluminium hydride, organolithium compounds (or Grignard reagents), or benzophenone ketyl. The last, used in a "THF still" is favoured by the present author. Various designs of THF still (which may be used or adapted for other solvents) are available. Two good ones are shown in Fig. 2.1. It is important that the THF used in the still should not be too wet; it should be predried with, for example, anhydrous magnesium sulphate followed by sodium wire. The predried solvent is placed in the flask A, together with benzophenone. Sodium or, better, potassium chips are added. When any initial effervescence has ceased the flask is heated, and the refluxing THF is returned to flask A via the three-way tap B. When the THF is dry, the intense blue colour of benzophenone ketyl develops, and the remaining pieces of sodium or potassium should have a bright surface. The tap B is then shut, and dry THF collects in the reservoir C. When reservoir C is full, condensate overflows and returns to the flask A. The second outlet from the tap B is connected to the apparatus to be used for the reaction, and the required volume of THF is transferred. Alternatively, small volumes of dry THF may be transferred by syringe via the septum D.

Of the many other ethers that have been or could be used as solvents for organolithium reactions, only DME warrants special mention. As a diether, its chelating properties can confer the same kind of activation as TMEDA, but to a lesser degree. It should be noted that aromatic ethers such as anisole are not usable as solvents for organolithium reagents, owing to their ready metallation (see Section 3.2).

Tertiary amines have been comparatively little used as solvents for organolithium compounds, though triethylamine probably warrants further trials. Pyridine cannot be used, as it is readily alkylated by organolithium compounds (see Section 5.2). TMEDA is more commonly used as an additive than a solvent: one molar equivalent is normally sufficient to exert the required activating effect, and an excess is susceptible to metallation (5).

*Hexametapol** is also attacked by organolithium compounds (6), but it is a particularly good promoter of single-electron-transfer reactions, so it is commonly used as a co-solvent (see below) rather than an additive. NB: *Hexametapol is a suspected carcinogen.*

Mixed solvent systems. It will be apparent from the discussion above that there are quite severe limitations on the use of certain solvents for organolithium compounds. Fortunately it is often possible to make use of the

---
*See note in List of Abbreviations (p. xvii).

desirable properties of these solvents, while minimizing their drawbacks, by using them in admixture with other solvents. For example, LDA is insoluble in hydrocarbons and rapidly cleaves THF, but may be stored for prolonged periods in THF–hydrocarbon mixtures (see p. 35). Similarly, hexametapol is useful for promoting conjugate addition of organolithium compounds to $\alpha\beta$-unsaturated carbonyl compounds, but as well as being liable to cleavage it freezes at 7°. However, hexametapol–THF mixtures can be used satisfactorily at low temperatures (7).

Mixed solvent systems are also the key to working at very low temperatures, where it is important not only that the reaction medium should remain liquid, but that its viscosity should remain low. A particularly useful combination is the "Trapp mixture", a 4 : 4 : 1 mixture of THF, diethyl ether and pentane (or hexane or light petroleum) (8).

2.2.2. Reactions at Low Temperatures

Reactions of organolithium compounds are often carried out at low temperatures—sometimes very low temperatures. Unfortunately, however, the published experimental details frequently give inadequate or inaccurate information on the true temperature at which a reaction takes place. In particular it is often stated that a reaction is carried out at $-78°$—the sublimation temperature of carbon dioxide at atmospheric pressure; this often means merely that the reaction vessel is immersed in a solid carbon dioxide–acetone bath, rather than that the internal temperature is $-78°$. In the experimental procedures recorded in this book, the statements in the primary literature have had perforce to be repeated, but with honourable exceptions they should be regarded with some reservations. The present author hopes that research workers, referees and editors will in future insist on *internal* temperatures being recorded.

TABLE 2.2

Solvents for Cooling Slush Baths

Solvent	Temperature of slush
Tetrachloromethane	$-23°$
Chlorobenzene	$-45°$
Chloroform	$-63°$
Ethyl acetate	$-84°$
Hexane	$-94°$
Methanol	$-98°$
Methylcyclohexane	$-126°$
Pentane	$-131°$

2.2. PRACTICAL CONSIDERATIONS

Besides solid carbon dioxide–acetone baths, and for temperatures below −78°, convenient cooling baths include solvents cooled to a slush by stirring with liquid nitrogen; a selection is listed in Table 2.2 (and see also Table 2.1 and Ref. (9)).

For those with generous equipment budgets, immersion coolers are now available giving temperatures as low as −100°.

2.2.3. Inert Atmospheres

For precise physico-chemical work, and for particular applications where traces of alkoxides may have significant effects, it may be necessary to employ vacuum-line techniques, rigorously purified atmospheres in glove boxes, and the like.* For most preparative applications, however, a sufficient degree of protection is achieved simply by maintaining a slight positive pressure of an inert gas over the reaction mixture. (Nevertheless, the present author would not recommend the over-relaxed attitude of a research student who, when caught in the act of pouring butyllithium from an open graduated cylinder, replied that approximately 100 litres of air would be required to react with one mole of butyllithium, so that the exposure he was giving his solution was insignificant!)

The cheapest form of inert atmosphere is "white spot" nitrogen, used directly from the cylinder without purification. While this is satisfactory for most experiments, it is preferable to use argon, particularly for preparations using lithium metal, whose surface becomes tarnished by a nitride film in contact with nitrogen.

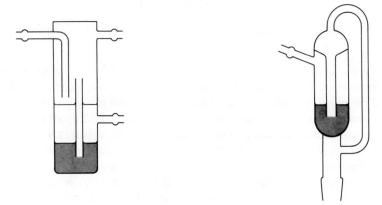

Fig. 2.2. Inert-gas bubblers.

* For information on such techniques see e.g. Refs (10) and (11).

10   2. GENERAL CONSIDERATIONS

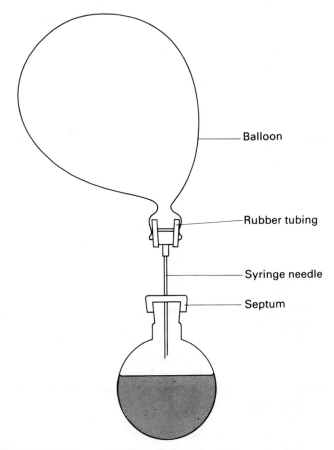

Fig. 2.3. Balloon method for maintaining an inert atmosphere.

To preserve an inert atmosphere in apparatus, it is best to maintain the internal pressure at slightly above atmospheric. An excess pressure of 1–2 cm of mineral oil is usually sufficient, and this is conveniently ensured by means of a bubbler such as those shown in Fig. 2.2, attached to the apparatus. This is designed in such a way that if the internal pressure should inadvertently fall below atmospheric (e.g. on rapid cooling), oil is not sucked into the apparatus and ingress of oxygen is minimized.

For small-scale work, a very simple device is a balloon connected to a syringe needle and inflated by the inert gas. The apparatus is flushed with the inert gas, the septum fitted, and the needle inserted, as shown in Fig. 2.3. However, the use of a balloon is *not* recommended for prolonged reactions, since oxygen can diffuse into the balloon to equilibrate the partial pressures.

2.2.4. Storage and Transfer of Organolithium Compounds; Safety Precautions

As organolithium compounds react vigorously, or sometimes violently,* with water and oxygen, they must be stored and transferred out of contact with both. As noted in the previous section, this is most easily accomplished by handling them under a blanket of nitrogen or argon. Commercial organolithium compounds are supplied in bulk in cylinders or tanks, and on a smaller scale in bottles fitted with sealed caps containing septa. In each case dead space in the container is filled with an inert gas. In order to transfer organolithium compounds from these containers (or from flasks in which they have been prepared in the laboratory) to reaction vessels or to smaller intermediate storage containers, use may be made of the pressure of the inert atmosphere. In the case of cylinders and tanks, manufacturers provide detailed procedures (see also Ref. (12)). On an intermediate scale, the vessels concerned are interconnected by a tube, which can be flushed with gas and then dipped below the surface of the solution, which is forced over by gas pressure or syphoned (Fig. 2.4). A very convenient form of connecting tube for fairly small quantities is a flexible double-ended hollow needle, which can be passed through septa at each end (Fig. 2.5).

For intermediate storage, modified Schlenk tubes (Fig. 2.6) may be used. The version with a burette side-arm is particularly convenient for dispensing a measured volume to a reaction vessel.

For small-scale transfer the syringe-and-septum method is routinely used. The following procedure is recommended.

(i) The syringe is flushed with inert gas.
(ii) If the organolithium solution is being withdrawn from a closed vessel, a positive internal pressure is maintained by passing inert gas through the septum. (Alternatively, the syringe is filled with inert gas, which is forced into the vessel before the solution is withdrawn.)
(iii) Slightly more than the required volume of solution is drawn into the syringe. The tip of the needle is then raised above the surface of the solution, and some inert gas is drawn into the syringe.
(iv) The syringe is withdrawn and immediately inverted, and the plunger is advanced until the barrel is just filled by the solution.

---

* The danger of auto-ignition of organolithium compounds on exposure to the atmosphere varies with the identity of the compound and with such factors as the nature of the solvent, the concentration and the humidity. Rigorous comparative studies have not been published, but rough guidelines to the degree of hazard for n-butyllithium in hydrocarbon solvents are summarized in Table 2.3; s- and t-butyllithium are even more hazardous.

TABLE 2.3

Pyrophoricity of n-Butyllithium in Hydrocarbon Solvents (*13*)

1. Concentrations of 15% or less of n-butyllithium in pentane, hexane, cyclohexane, or heptane are considered flammable and of low degree of pyrophoricity. Most commercial applications use 15% solutions for reduced hazard in handling.
2. Usually, but not always, solutions at 20% concentration or less will not ignite immediately and spontaneously at normal room temperature and relative humidities below 70%.
3. Concentrations above 25% are usually pyrophoric under any practical range of humidity.
4. Concentrations in the range of 50–80% are considered most hazardous, with a peak in pyrophoricity in the range 60–75%. Such solutions ignite immediately on exposure to air.
5. From 80 to 92% the probability of immediate spontaneous ignition tends to decrease slightly owing to the decreasing amount of solvent that is considered supportive "flammable fuel".
6. Increase in relative humidity increases the probability for spontaneous ignition.
7. Increase in the flashpoint of the hydrocarbon solvent tends to decrease the probability of spontaneous ignition. For example flashpoints of common solvents increase as follows:

n-pentane	$-40°$
n-hexane	$-22°$
cyclohexane	$-17°$
n-heptane	$-4°$

8. Highly volatile solvents will evaporate quickly, resulting in concentration of the non-volatile n-butyllithium and increasing the degree of pyrophoricity.
9. The following conditions increase the degree of pyrophoricity (probability of ignition) significantly for all concentrations:
    (*a*) contact with water or moist materials;
    (*b*) contact with reactive or combustible materials (paper tissues must *not* be used to wipe up spillages!);
    (*c*) local overheating and ignition of solvent vapours.

(v) The required amount of solution is dispensed into the receiver.
(vi) The excess of solution is discharged into a quenching medium (e.g. an acetone (or ethyl acetate)–hydrocarbon mixture), and the syringe is immediately flushed out, disassembled, cleaned, and dried. (Lithium hydroxide and alkoxides can rapidly "seize" syringes, taps, stoppers, etc.)

Despite the greatest care in keeping and using organolithium compounds under safe conditions, accidents may happen. Appropriate precautions must

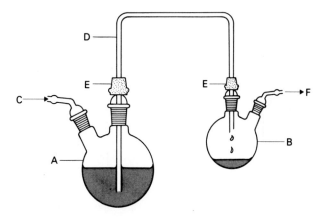

Fig. 2.4. Transfer of organolithium solution under inert atmosphere: A, storage flask; B, receiving flask; C, inert gas in; D, transfer tube; E, sleeve; F, to bubbler.

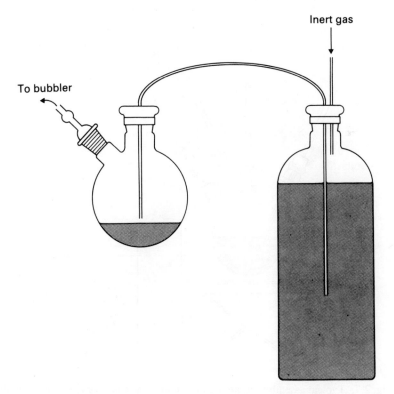

Fig. 2.5. Transfer by means of a double-ended needle.

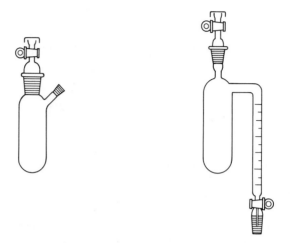

Fig. 2.6. Modified Schlenk tubes.

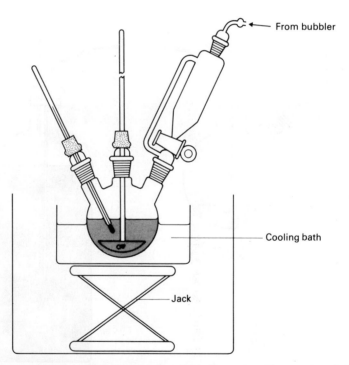

Fig. 2.7. Typical assembly for an organolithium reaction. For reactions in boiling solvent the thermometer is replaced by a condenser.

2.2. PRACTICAL CONSIDERATIONS

therefore always be taken. Personal protective clothing and safety spectacles should be worn. Experiments should be conducted behind screens. Dishes to catch spillages should be provided. In case of fire, EXTINGUISHERS USING WATER OR HALOGENATED HYDROCARBONS MUST NOT BE USED. For small laboratory-scale fires carbon dioxide extinguishers may be usable, but the most generally satisfactory extinguishers for organolithium fires are the dry-powder type.

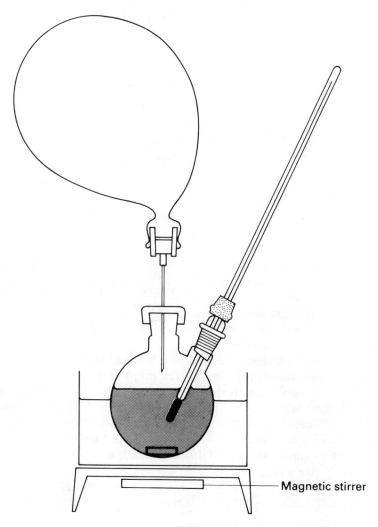

Fig. 2.8. Typical assembly for a small-scale organolithium reaction.

2.2.5. Typical Assemblies for Reactions

Two typical assemblies of apparatus are shown in Figs 2.7 and 2.8. Assemblies of these types can be readily adapted to accommodate variations in procedures and the particular models of equipment available.

2.3. DETECTION AND ESTIMATION OF ORGANOLITHIUM COMPOUNDS

2.3.1. Detection

Gilman Colour Tests

For detecting the presence of an organolithium compound (for example for checking whether any remains at the finish of a reaction) Gilman's Colour Test 1 is convenient:

$$(4\text{-}Me_2NC_6H_4)_2C=O + RLi \longrightarrow (4\text{-}Me_2NC_6H_4)_2C(R)OLi$$

$$\xrightarrow[\text{(ii) } I_2,\text{ AcOH}]{\text{(i) } H_2O} [(4\text{-}Me_2NC_6H_4)_2CR]^{\oplus} \ I^{\ominus}$$

A sample (0.5–1 ml) of the test solution is added to an equal volume of a 1% solution of Michler's ketone, 4,4′-bis(dimethylamino)benzophenone, in dry benzene. A few drops of water are added, and the mixture is shaken.* A few drops of an 0.2% solution of iodine in glacial acetic acid are added, resulting in the formation of a blue or green colour.

Gaidis proposed two modifications, to distinguish between alkyl- and aryllithium compounds (*15*).

(a) Following hydrolysis, the solution is buffered to pH 9 by the addition of 20% aqueous catechol (1–2 ml). A few drops of 0.5% iodine in benzene are added and the solution is shaken. A green to blue colour in the organic layer indicates the presence of an aryllithium compound. The mixture is then acidified with acetic acid to reveal the presence of an alkyllithium compound.

(b) Following hydrolysis the solution is acidified with acetic acid (5–15 drops). 20% aqueous sodium bisulphite is added; aryl- but not alkyllithium compounds give a green colour.

Gilman's Colour Test 2 (*16*) which also distinguishes alkyl- from aryllithium compounds, depends on the rapid metal–halogen exchange of the

---

* Aerial oxidation sometimes leads to the formation of the blue-green colour at this stage.

2.3. DETECTION AND ESTIMATION

former with 4-bromo-N,N-dimethylaniline:

$$4\text{-BrC}_6\text{H}_4\text{NMe}_2 + \text{RLi} \rightleftharpoons 4\text{-LiC}_6\text{H}_4\text{NMe}_2 + \text{RBr}$$

$$4\text{-LiC}_6\text{H}_4\text{NMe}_2 \xrightarrow[\text{(ii) H}_3\text{O}^+]{\text{(i) Ph}_2\text{CO}} [\text{Ph}_2\text{CC}_6\text{H}_4\text{NMe}_2]^\oplus$$

The solution to be tested (0.5–1 ml) is added to an equal volume of a 15% solution of 4-bromo-N,N-dimethylaniline in dry benzene. A 15% solution of benzophenone in dry benzene (1 ml) is added. After a few seconds, water is added and then the mixture is acidified with concentrated hydrochloric acid. A positive result is indicated by a red colour in the aqueous layer.

2.3.2. Estimation

Note. High-precision analyses involving lithium or organolithium compounds are not normally required for preparative work, and in any case volumetric methods are normally used. However, it should be noted that the atomic weight of commercially supplied lithium may differ appreciably from the "natural-abundance" value of 6.94, owing to the selective removal of one isotope (usually ^{6}Li) for nuclear energy purposes (*17*). Commercial lithium may also contain a variable proportion of sodium.

For routine, non-critical estimations of freshly prepared solutions of organolithium compounds in hydrocarbon solvents, an aliquot of the test solution may be added to water and then titrated with standard acid. However, this simple determination of total base will give a high estimate if lithium hydroxide (from hydrolysis of the organolithium compound) or alkoxide (from oxidation or from cleavage of an ethereal solvent) are present.

The best-established methods for estimating organolithium compounds in the presence of lithium hydroxide or alkoxides are the Gilman double titration and its variants. In these, total base is first determined. An aliquot of the test solution is then treated with an organic halide of sufficient reactivity to convert the organolithium compound into non-basic lithium halide while leaving any hydroxide or alkoxide unchanged for a second titration. Gilman originally used benzyl chloride (*18*), but 1,2-dibromoethane and allyl bromide have been reported to be more generally satisfactory (*19, 20*). Conditions can be established that give precise, reproducible results in particular cases; the ASTM method for assaying n-butyllithium in hydrocarbon solutions is an example (*21*).

Estimation of n-butyllithium in hydrocarbons by Gilman double titration

This procedure is based on Ref. (*21*), which gives a more detailed *modus operandi*.

(a) An aliquot of the test solution (3.0 ml) is added by syringe to hexane (20 ml) in a 250 ml conical flask. (If greater precision than that afforded by a graduated syringe is required, the amount of sample is measured by weighing the syringe before and after delivery.) Water (35 ml) is added, dropwise at first, with swirling. The mixture is titrated with 0.1 M HCl to pH 7.

(b) A dry container (flask or serum bottle) is flushed with nitrogen and closed with a serum cap. Benzyl chloride (4 ml) and dry diethyl ether (10 ml) are injected by syringe. An aliquot of the test solution (4.0 ml) is added by syringe. The container is shaken, then allowed to stand for a minute or so. The closure is removed, and it and the neck of the container are rinsed into the container with water from a wash bottle. Water is added to a total of *ca* 40 ml. The mixture is titrated to a pink endpoint (methyl orange) with 0.1 M HCl.

Because of the difficulty of establishing standard conditions for the double titration procedure, applicable to a variety of organolithium compounds and solvents (see Ref. (*22*) for a detailed review), many attempts have been made to devise alternative methods, not subject to interference by other bases, using a range of volumetric, gravimetric, spectroscopic, and thermometric techniques (see Ref. (*22*) and General Ref. A). Of these, several simple titrimetric procedures have been found to be reasonably general, and adequate for most routine use. Examples are listed in Table 2.4. All the indicators listed are commercially available.

TABLE 2.4

Direct Titration of Organolithium Compounds

Indicator	Titrant	Notes	Ref.
Triphenylmethane in DMSO	Benzoic acid		(*23*)
1,10-Phenanthroline or 2,2'-biquinoline	Butan-2-ol	(*a*)	(*24*)
Diphenylacetic acid	(*b*)		(*25*)
Biphenyl-4-ylmethanol	(*b*)		(*26*)
2,5-Dimethoxybenzyl alcohol	(*b*)		(*27*)
1,3-Diphenylacetone tosylhydrazone	(*b*)		(*28*)
N-Phenyl-1-naphthylamine	Butan-2-ol	(*c*)	(*29*)
N-Benzylidenebenzylamine	Butan-2-ol	(*c*)	(*30*)

(*a*) Ethers interfere.
(*b*) The indicator is also the titrant.
(*c*) Back-titration, after reaction of the test solution with an excess of the indicator.

REFERENCES

1. W. N. Setzer and P. v. R. Schleyer, *Adv. Organomet. Chem.* **24**, 353 (1985).
2. D. Seebach, Proceedings of the R. A. Welch Foundation Conferences on Chemical Research XXVII, Houston, 1984.
3. P. v. R. Schleyer, *Pure Appl. Chem.* **56**, 151 (1984).
4. W. Bauer and D. Seebach, *Helv. Chim. Acta* **67**, 1972 (1984).
5. D. J. Peterson, *J. Organomet. Chem.* **9**, 373 (1967).
6. E. M. Kaiser, J. D. Petty and L. E. Solter, *J. Organomet. Chem.* **61**, C1 (1973).
7. W. C. Still and A. Mitra, *Tetrahedron Lett.* 2659 (1978); see also General Ref. D(ii), p. 28.
8. G. Köbrich, *Angew. Chem. Int. Ed. Engl.* **6**, 41 (1967).
9. A. J. Gordon and R. A. Ford, *The Chemist's Companion*, Wiley, New York, 1972.
10. D. F. Shriver and M. A. Drezdzon, *The Manipulation of Air-Sensitive Compounds*, 2nd edn, Wiley, New York, 1986; see also G. B. Gill and D. A. Whiting, *Aldrichimica Acta* **19**, 31 (1986).
11. J. J. Eisch, *Organometallic Syntheses*, Vol. 2, Wiley, New York, 1981.
12. R. Anderson, *Chem. Ind. (London)* 205 (1984).
13. Lithium Corporation of America, Product Bulletin on n-Butyllithium.
14. H. Gilman and F. Schulze, *J. Am. Chem. Soc.* **47**, 2002 (1925).
15. J. M. Gaidis, *J. Organomet. Chem.* **8**, 385 (1967).
16. H. Gilman and J. Swiss, *J. Am. Chem. Soc.* **62**, 1847 (1940).
17. N. N. Greenwood and A. Earnshaw, *Chemistry of the Elements*, Pergamon, Oxford, 1984.
18. H. Gilman and A. H. Haubein, *J. Am. Chem. Soc.* **66**, 1515 (1944).
19. H. Gilman and F. K. Cartledge, *J. Organomet. Chem.* **2**, 447 (1964).
20. R. R. Turner, A. G. Altenau and T. C. Cheng, *Analyt. Chem.* **42**, 1835 (1970).
21. American Society for Testing and Materials, ASTM Designation E233-68.
22. T. R. Crompton, *The Chemistry of the Metal–Carbon Bond*, ed. F. R. Harley and S. Patai, Vol. 1, Chap. 15, Wiley, Chichester, 1982.
23. R. L. Eppley and J. A. Dixon, *J. Organomet. Chem.* **8**, 176 (1967).
24. S. C. Watson and J. F. Eastham, *J. Organomet. Chem.* **9**, 165 (1967).
25. W. G. Kofron and L. M. Baclawski, *J. Org. Chem.* **41**, 1879 (1976).
26. E. Juaristi, A. Martinez-Richa, A. Garcia-Rivera and J. S. Cruz-Sanchez, *J. Org. Chem.* **48**, 2603 (1983).
27. M. R. Winkle, J. M. Lansinger and R. C. Ronald, *J. Chem. Soc. Chem. Commun.* 87 (1980).
28. M. F. Lipton, C. M. Sorensen, A. C. Sadler and R. H. Shapiro, *J. Organomet. Chem.* **186**, 155 (1980).
29. D. E. Bergbreiter and E. Pendergrass, *J. Org. Chem.* **46**, 219 (1981).
30. L. Duhamel and J.-C. Plaquevent, *J. Org. Chem.* **44**, 3404 (1979).

—3—
Preparation of Organolithium Compounds

Several organolithium reagents are made commercially, some of them on a considerable scale;* n- and s-butyllithium in hydrocarbon solvents, for example, are sold in tonnage quantities. The majority are prepared by the reaction of organic halides with lithium metal (Section 3.1.1), but metal-halogen exchange (3.1.3) and metallation (3.2) are also used. They are normally sold as solutions in hydrocarbons or ethers.

The list of organolithium compounds on sale as laboratory reagents varies from time to time, but some, offered in 1986–87, are shown in Table 3.1.

3.1. PREPARATION FROM ORGANIC HALIDES

3.1.1. By Reaction with Lithium Metal

$$RX + 2Li \longrightarrow RLi + LiX$$

Since organolithium compounds are made industrially by the reaction of organic halides with lithium, much effort has been expended on optimizing the yields of such reactions. The best conditions vary for specific compounds, but some generalizations may be made.

(a) Although the reactivity of organic halides towards lithium is in the order RI > RBr > RCl (the fluorides rarely react at all), their susceptibility towards Wurtz-type reactions is in the same order. Moreover, lithium iodide and lithium bromide are more soluble than lithium chloride in ethers. Thus, for highest yields and for halide-free solutions, organic chlorides are preferred. When initiation of the reaction is difficult, or when the presence

---

Large-scale Suppliers of Organolithium Compounds (in alphabetical order): Chemetall GmbH, Lithium Department, Reuterweg 14, D6000 Frankfurt 1, Federal Republic of Germany; Foote Mineral Company, Route 100, Exton, Pennsylvania 19341, USA; Lithium Corporation of America, 449 North Cox Road, Gastonia, North Carolina 28054, USA; Lithium Corporation of Europe, Commercial Road, Bromborough, Merseyside L62 3NL, UK.

TABLE 3.1
Organolithium Compounds Sold as Laboratory Reagents

Organolithium compound	Solvent	Concentration	Supplier
Methyllithium (low in halide)	Diethyl ether	1.4 M	(a, b, c)
Methyllithium–lithium bromide complex	Diethyl ether	2 M	(a, b)
Allyllithium	Diethyl ether	1 M	(b)
n-Butyllithium	Hexanes	1.6, 2.5 and 10 M	(a, b, c)
s-Butyllithium	Cyclohexane	1.3 M	(a, b)
t-Butyllithium	Pentane	1.7 M	(a, b, c)
Neopentyllithium	(d)	—	(b)
Phenyllithium	Cyclohexane–diethyl ether (70:30)	2 M	(a)
Phenyllithium–lithium bromide complex	Diethyl ether	1 M	(a)
Phenyllithium–TMEDA	Benzene	ca. 1 M	(b)
4-Tolyllithium	(d)	—	(b)
Trimethylsilylmethyllithium	Pentane	1.0 M	(a)
(Diphenylphosphino)methyllithium–TMEDA complex	(d)	—	(b)

(a) Adrich Chemical Company, P.O. Box 355, Milwaukee, Wisconsin 53201, USA, and The Old Brickyard, New Road, Gillingham, Dorset SP8 4JL, UK.
(b) Ventron Alfa Products, 150 Andover Street, Danvers, Massachusetts, USA: Zeppelinstrasse 7, Postfach 6540, D-7500 Karlsruhe 1, FRG; University of Warwick Science Park, Sir William Lyons Road, Coventry CV4 7EZ, UK.
(c) Fluka AG, CH-9470 Buchs, Switzerland.
(d) Solid.

of lithium halide in the solution is desired, bromides or iodides may be used. It should be noted that even allyl and benzyl chlorides undergo Wurtz-type reactions so readily that their reaction with lithium metal is not a useful method for preparing allyl- and benzyllithium compounds.

(b) The lithium used should be finely divided and clean. For easily prepared reagents, lithium metal in rod form may be washed free from mineral oil, beaten into a sheet, and cut with strong scissors so that chips fall directly into the reaction solvent. Lithium is too hard to be extruded as wire by a standard sodium press, but a strong press with a wider-bore die may be used. Lithium metal is now commercially available as wire and shot and as a dispersion in mineral oil. Preparation of a dispersion in the laboratory is somewhat hazardous, as it involves shaking or stirring the molten metal in a high-boiling inert medium (1, 2). As noted in Section 2.2.3, it is desirable to use

3.1. FROM ORGANIC HALIDES

an atmosphere of argon rather than nitrogen in preparations requiring a clean metal surface.

(c) The presence of a small amount of sodium in lithium increases its reactivity toward organic halides (1, 3). For particular purposes (for example in the preparation of certain polymerization initiators) the presence of any sodium is undesirable, but in most cases 1–2% sodium aids initiation while having little effect on the yield or on the properties of the product. Higher concentrations of sodium may lead to loss of yield through increasing side-reactions (particularly the Wurtz reaction).

(d) The halide should be added slowly, with efficient stirring. There is usually an induction period, and only a small proportion of the halide should be added before the reaction starts. Otherwise not only will the yield be reduced but there will be the risk of a runaway exothermic reaction (3a).

It has been reported that ultrasonic irradiation may aid the initiation of the reactions, and in some cases increase their yields (4).

It should be emphasized that, although for highest yields (and in some cases for any yield at all) great attention to details is required, quite acceptable results can often be obtained by straightforward Grignard-type procedures; a simple preparation of methyllithium on a three-mole scale is described in *Organic Syntheses* (5). Some representative examples follow, and a selection of others is listed in Table 3.2.

t-Butyllithium

The preparation of t-butyllithium by the reaction of lithium with t-butyl chloride is particularly difficult since t-butyllithium and t-butyl chloride react together readily to give isobutane and isobutene. The following procedure is that described by Smith (1). The procedures of Kamienski and Esmay (3) and Giancaspro and Sleiter (6) employ similar principles.

A three-necked round-bottomed flask is fitted with a high-speed stirrer, pressure-equalizing dropping funnel and reflux condenser, and flushed with argon. Dry pentane (470 ml) and 1% sodium–lithium alloy dispersion are added. The mixture is heated to reflux and stirred vigorously, and the addition of t-butyl chloride (107 ml) containing t-butanol (0.5 ml) is begun. Reaction normally begins after the addition of 3–5 ml; no more than 12 ml should be added until the reaction has initiated. When the reaction has started, the remainder of the chloride is added during 3 h at such a rate that a gentle reflux is maintained without external heating. Stirring is continued for 2 h, as the mixture slowly cools. The resulting slurry is filtered under argon pressure through a sintered-glass (4–8 μm) filter. The reaction flask is rinsed with dry pentane (200 ml), which is then poured through the filter cake. (DANGER: the washed filter cake is pyrophoric.) The combined pentane solutions (475 g) contain 10.7% t-butyllithium (by Gilman titration), representing a total yield of 80%.

Phenyllithium

In contrast with t-butyllithium, the preparation of phenyllithium is straightforward and gives high yields with minimal precautions. Phenyllithium is insoluble in hydrocarbons, but cleaves ethers only slowly, so it is conveniently prepared in diethyl ether, as described by Gilman and Morton (7).

Dry ether (500 ml) is placed in a 2 l three-necked flask equipped with a stirrer, a reflux condenser carrying a nitrogen inlet, and a solids funnel, and flushed with nitrogen. A flow of nitrogen is maintained as lithium (29.4 g) is cut into small pieces, which are allowed to fall into the flask. The solids funnel is replaced by a dropping funnel containing a solution of bromobenzene (314 g) in dry ether (1 l). The stirrer is started and about 40 drops of the bromobenzene solution are run into the flask. When the reaction begins (indicated by the contents of the flask becoming slightly cloudy), more bromobenzene is added until vigorous boiling begins. The flask is cooled in an ice bath, and the addition of bromobenzene is continued at such a rate as to maintain steady refluxing.[a] After most of the bromobenzene has been added, the flask is removed and stirring is continued until refluxing stops.[b] The mixture is decanted and/or filtered under nitrogen. The yield of such preparations, determined by simple titration, is 95–99%.

[a] If refluxing stops, the ice bath should be removed until the exothermic reaction recommences. In small-scale preparations cooling is usually unnecessary, but it is advisable to have an ice bath at hand as a safety precaution.
[b] The preparation to this stage typically takes about 2 h.

Ethyllithium

Ethyllithium is only sparingly soluble in pentane, and cleaves ethers fairly rapidly. The procedure described by Bryce-Smith and Turner gives a high yield in suspension in pentane (8). If a solution in a hydrocarbon is required, and the hazard can be tolerated, benzene may be used. If a halide-free solution is required, the very finely divided precipitate is very difficult to remove by filtration, so centrifugation may be necessary (10).

Lithium (2.0 g) is extruded as wire of 0.5 mm diameter directly into pentane (40 ml) under an atmosphere of nitrogen. The mixture is stirred vigorously and heated under reflux as a mixture of bromoethane (13.1 g) with

TABLE 3.2
Preparation of Organolithium Compounds from Organic Halides and Lithium Metal

Organic halide	Solvent	Yield of corresponding organolithium compound (%)	Ref.
MeCl	Et$_2$O	70–89^a	(11)
MeI	Et$_2$O	82^b	(12)^c
BuCl	Pentane	93–98	(8^c, 13)
BuBr	Et$_2$O	80–90	(14)
Me$_3$CCH$_2$Cl	n-Pentane	90	(15)
	Et$_2$O	85	(16)
cyclo-C$_3$H$_5$Br	Et$_2$O	88	(17)
cyclo-C$_6$H$_{11}$Cl	Light petroleum	>70	(18)
C$_{10}$H$_{15}$Cl^d	Pentane	82	(19)^c
Me$_2$N(CH$_2$)$_3$Cl	THF	>90	(20)
CH$_2$=CHCl	THF	69–79	(21)
Me$_2$CH=C(Br)(Me)	Et$_2$O	73–75	(22)
C$_6$H$_9$Cl^e	Et$_2$O	65	(23)
C$_{10}$H$_{16}$Cl^f	Et$_2$O	>60	(24)
Me$_3$SiCH$_2$Cl	Et$_2$O or hexane	90	(25)

^{a} Halide free. ^{b} Lithium iodide complex.
^{c} General procedure described, applicable to several other examples.
^{d} 1-Chloroadamantane. ^{e} 1-Chlorocyclohexene. ^{f} 2-Chlorobornane.

pentane (40 ml) is added dropwise during 3–4 h. Stirring under reflux is continued for a further 1 h. The yield of ethyllithium, as determined by analysis of aliquots of the stirred suspension, exceeds 90%.

References

1. W. N. Smith, *J. Organomet. Chem.* **82**, 1 (1974).
2. D. L. Perrine and H. Rapoport, *Analyt. Chem.* **20**, 635 (1948).
3. C. W. Kamienski and D. L. Esmay, *J. Org. Chem.* **25**, 1807 (1960).
3a. D. Service, *Chem. Br.* **23**, 27 (1987); C. W. Kamienski, *Ibid.* p. 531.
4. J.-L. Luche and J.-C. Damiano, *J. Am. Chem. Soc.* **102**, 7926 (1980).
5. U. Schöllkopf, J. Paust and M. R. Patsch, *Org. Synth.*, Collective Vol. V, 860 (1973).
6. C. Giancaspro and G. Sleiter, *J. Prakt. Chem.* **321**, 876 (1979).
7. H. Gilman and J. W. Morton, *Org. React.* **8**, 286 (1954).
8. D. Bryce-Smith and E. E. Turner, *J. Chem. Soc.* 861 (1953).
9. M. Kaspar and J. Trekoval, *Coll. Czech. Chem. Commun.* **42**, 1969 (1977).
10. S. W. Simpson, Personal communication.

11. M. J. Lusch, W. V. Phillips, R. F. Sieloff, G. S. Nomura and H. O. House, *Org. Synth.* **62**, 101 (1984).
12. H. Gilman, E. A. Zoellner and W. M. Selby, *J. Am. Chem. Soc.* **55**, 1252 (1933).
13. E. H. Amonoo-Neizer, R. A. Shaw, D. O. Skovlin and B. C. Smith, *Inorg. Synth.* **8**, 19 (1966).
14. R. G. Jones and H. Gilman, *Org. React.* **6**, 339 (1951).
15. B.-H. Chang, H.-S. Tung and C. H. Brubaker, *Inorg. Chim. Acta* **51**, 143 (1981).
16. H. D. Zook, J. March and D. F. Smith, *J. Am. Chem. Soc.* **81**, 1617 (1959).
17. D. Seyferth and H. M. Cohen, *J. Organomet. Chem.* **1**, 15 (1963).
18. O. H. Johnson and W. H. Nebergall, *J. Am. Chem. Soc.* **71**, 1720 (1949).
19. G. Molle, P. Bauer and J. E. Dubois, *J. Org. Chem.* **48**, 2975 (1983).
20. C. D. Eisenbach, H. Schnecko and W. Kern, *Eur. Polym. J.* **11**, 699 (1975).
21. W. N. Smith, *J. Organomet. Chem.* **82**, 7 (1974).
22. R. S. Threlkel, J. E. Bercaw, P. F. Seidler, J. M. Stryker and R. D. Bergman, *Org. Synth.* **65**, 42 (1987).
23. E. A. Braude and J. A. Coles, *J. Chem. Soc.* 2014 (1950).
24. G. W. Erickson and J. L. Fry, *J. Org. Chem.* **52**, 462 (1987).
25. C. Tessier-Youngs and O. T. Beachley, *Inorg. Synth.* **24**, 95 (1986).

3.1.2. By Reaction with Lithium Salts of Radical Anions

The reaction of organic halides with lithium metal, leading to organolithium compounds, almost certainly involves electron transfer (*1*), so it is not unexpected that under appropriate conditions electron transfer from a radical anion to an organic halide, with lithium as counter-ion, may also give organolithium compounds. With lithium naphthalene (*2*) yields are variable, but can be high with aryl chlorides (*3, 4*). With lithium 4,4'-di-t-butylbiphenyl excellent yields are reported in several cases (*5, 6*), and this reagent gives some organolithium compounds that are unobtainable by other methods (*7, 8*).

Lithium 4,4'-di-t-butylbiphenyl (LiDBB) (5, 6, 9)

4,4'-Di-t-butylbiphenyl (5.0 g, 18.8 mmol)^a is placed in a dry 100 ml three-necked flask, flushed with argon,^{b} and equipped with a magnetic stirrer. Dry THF (62 ml) is added.^{c} Lithium (0.11 g, 15.9 mmol) is cut into slivers in a rapid stream of argon emerging from the flask, the slivers being allowed to drop into the THF. After a few minutes an intense deep blue colour develops. The mixture is cooled to 0° and stirred rapidly for about 3 h.

^{a} It is essential to use pure 4,4'-di-t-butylbiphenyl. It may be prepared as described by Horne (*10*) and freed from traces of iron compounds by chromatography on basic alumina.
^{b} If nitrogen is used, the nitride film on the lithium inhibits the reaction.

^{c} No radical-anion formation occurs unless the apparatus and solvent are scrupulously dried. It has been reported that initiation of reactions of lithium with aromatic hydrocarbons is promoted by ultrasonic irradiation (*11*).

7-Lithionorbornadiene (*7, 9*)

The solution of LiDBB is cooled to $-78°$ and a solution of 7-chloronorbornadiene (0.67 g, 5.3 mmol) (*12*) in dry THF (0.5 ml) is added by means of a syringe. After a few minutes the blue colour is discharged.

Subsequent reactions of the solution with deuterium oxide, carbon dioxide or DMF give the appropriate 7-substituted norbornadienes in isolated yields of up to 64%.

References

1. J.-E. Dubois, P. Bauer and B. Kaddani, *Tetrahedron Lett.* **26**, 57 (1985).
2. C. G. Screttas and M. Micha-Screttas, *J. Org. Chem.* **43**, 1064 (1968).
3. C. G. Screttas, *J. Chem. Soc. Chem. Commun.* 752 (1972).
4. J. Barlengua, J. Florez and M. Yus, *J. Chem. Soc. Perkin Trans. 1*, 3019 (1983).
5. P. K. Freeman and L. L. Hutchinson, *Tetrahedron Lett.* 1849 (1976).
6. P. K. Freeman and L. L. Hutchinson, *J. Org. Chem.* **45**, 1924 (1980).
7. J. Stapersma and G. W. Klumpp, *Tetrahedron* **37**, 187 (1981).
8. J. Stapersma, P. Kuipers and G. W. Klumpp, *Rec. Trav. Chim. Pays Bas* **101**, 213 (1982).
9. F. Binns, Personal communication.
10. D. A. Horne, *J. Chem. Educ.* **60**, 246 (1983).
11. P. Boudjouk, R. Sooriyakumaran and B. H.-Han, *J. Org. Chem.* **51**, 2818 (1986).
12. S. C. Clark and B. C. Johnson, *Tetrahedron* **24**, 5067 (1968).

3.1.3. By Metal–Halogen Exchange

The general reaction,

$$\text{RLi} + \text{R'X} \rightleftharpoons \text{R'Li} + \text{RX}$$

has features that make it extremely valuable for preparing organolithium compounds. The equilibrium lies towards the side giving the organolithium compound whose organic group is better able to accommodate partial carbanionic character; it is thus particularly useful for preparing aryl- and 1-alkenyllithium compounds by reaction of butyllithium with aryl and alkenyl halides. With iodo- and bromo-compounds the reaction is general, and often proceeds remarkably rapidly even at very low temperatures. It is less commonly satisfactory with chloro-compounds (polychloroaromatic compounds being a notable exception (*1*)), and does not normally occur with fluoro-compounds. The reaction is normally carried out in ether solvents, though it does take place (but more slowly) in hydrocarbons.

TABLE 3.3

Lithium–Halogen Exchange in the Presence
of Reactive Functional Groups (5)

Functional group	Ref.
Ester[a]	(6)
Carboxylate	(7, 8)
Carboxamide	(8)
Nitrile	(9)
Epoxide	(10)
Formamido	(11)
Nitro	(12)

[a] Only when sterically hindered.

Because metal–halogen exchange proceeds rapidly under mild conditions, potential side-reactions such as alkylation of, or elimination from, the organic halide are not usually troublesome. Moreover, the reaction often occurs at such low temperatures that even α-halogenoalkyllithium compounds (carbenoids) and o-halogenoaryllithium compounds (arynoids) are stable. It can even be carried out in the presence of functional groups normally regarded as incompatible with organolithium reagents: some examples are listed in Table 3.3.

Remarkably, the metal–halogen exchange reaction can compete with reactions of organolithium compounds with dimethyl sulphate (2) and with their protonation by water (tritium oxide) (3) or even an intramolecular carboxylic acid group (4).

One feature of the reaction that can cause complications is the presence of the alkyl halide product. When the desired organolithium reagent is warmed for subsequent reaction it can couple with the alkyl halide, giving R—R'. Alternatively, the alkyl halide can react with a product. For example if a solution of 3-lithio-2-phenylbenzo[b]thiophene, prepared by the method described below for the parent compound, is allowed to warm to room temperature, the major product is the butylthio compound [1], formed as follows (13):

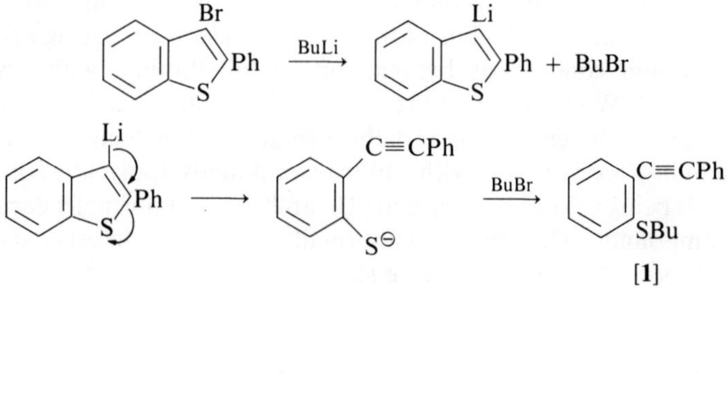

3.1. FROM ORGANIC HALIDES

This type of side-reaction may be avoided by the use of two equivalents of t-butyllithium (*14, 15*). The second equivalent rapidly reacts with the t-butyl halide formed to give the innocuous by-products lithium halide and isobutene.

The early work on the metal–halogen exchange reaction has been well reviewed, with a selection of experimental procedures (*16*). Extensive tables are given in General Ref. A, and a selection of later examples is listed in Table 3.4.

1-Hexenyllithium

This procedure (*15*) illustrates the use of t-butyllithium for obtaining solutions of organolithium compounds free from alkyl halides, and also the use of very low temperatures.

1-Bromohex-1-ene (1.63 g, 10 mmol) and Trapp mixture[a] (24 ml) are cooled to $-120°$ by means of a light-petroleum–acetone–isopropanol–liquid-nitrogen slush bath, and stirred as t-butyllithium (20 mmol) in pentane is added dropwise during 20 min. The resulting yellow solution is stirred at between $-120°$ and $-110°$ for 2 h. Solutions thus prepared have given products from reactions with electrophiles in yields greater than 90% (see p. 137).

Another example of the preparation of a 1-alkenyllithium compound by metal–halogen exchange is described on p. 113.

[a] See p. 8.

Pentachlorophenyllithium

This procedure, which gives a useful precursor for tetrachlorobenzyne (see p. 168) as well as an intermediate for synthesizing a variety of organic and organometallic pentachlorophenyl derivatives (*1*) is based on that first reported by Rausch *et al.* (*17*) and elaborated in *Organic Syntheses* (*18*). It may be readily adapted to a smaller scale.

A 5 l three-necked flask is equipped with a stirrer, a low-temperature thermometer and a 500 ml pressure-equalizing dropping funnel (so placed that the drops fall directly into the solution) bearing a gas inlet tube at its top. Hexachlorobenzene (28.5 g) is placed in the flask, the apparatus is flushed with nitrogen and diethyl ether (600 ml) is added. The resulting suspension is stirred as it is cooled below $-70°$. A solution of n-butyllithium (0.110 mol) in hexane is added to the stirred suspension during 30 min, the temperature being kept below $-70°$. Stirring is continued for a further 90 min, as the temperature is allowed to rise to *ca.* $-60°$.

TABLE 3.4

Preparation of Organolithium Compounds by Metal–Halogen Exchange

Organic halide	Metallating reagent (solvent)	Product (yield %)	Ref.
CH_3I [a]	BuLi (Et_2O or THF)	CH_3Li (quant.) [a,b]	(21)
$BuCHBr_2$	BuLi (THF–Et_2O–pentane) [c]	BuCHBrLi (64)	(22)
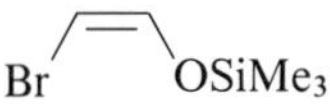 Br–CH=CH–$OSiMe_3$	Bu^tLi (Et_2O or THF)	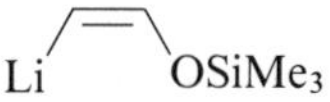 Li–CH=CH–$OSiMe_3$ (84)	(23)
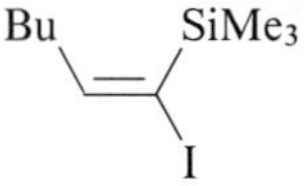 Bu–C(I)=C(SiMe_3)–	Bu^nLi (Et_2O–hexane)	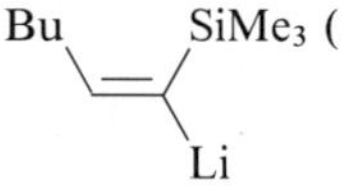 Bu–C(Li)=C($SiMe_3$)– (>90)	(24)
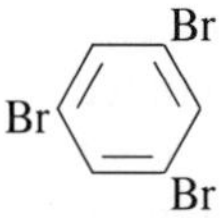 1,2,4-tribromobenzene (Br, Br, Br)	BuLi (Et_2O)	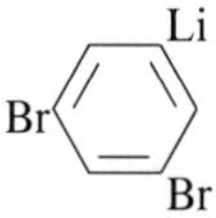 2,5-dibromo-Li-benzene (88) [d]	(25)
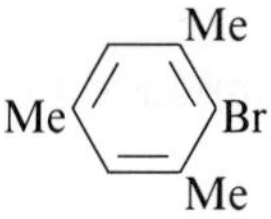 2,4,6-trimethylbromobenzene	BuLi (hexane)	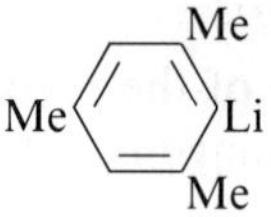 2,4,6-trimethylphenyllithium (quant.)	(26)
2-$MeSC_6H_4Br$	BuLi (Et_2O)	2-$MeSC_6H_4Li$ (89)	(27)
4-$Me_3SiC_6H_4Br$	BuLi (Et_2O)	4-$Me_3SiC_6H_4Li$ (67)	(28)
C_6F_5Br	BuLi (Et_2O)	C_6F_5Li (83)	(29)
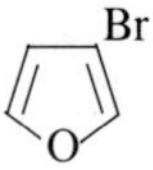 2-bromofuran	BuLi (Et_2O or THF)	 2-lithiofuran (97)	(30)
 N-($SiPr^i_3$)-2-bromopyrrole	BuLi (THF)	 N-($SiPr^i_3$)-2-lithiopyrrole (92)	(31)
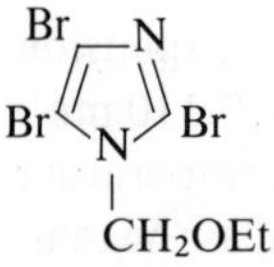 N-(CH_2OEt)-2,4,5-tribromoimidazole	BuLi (Et_2O)	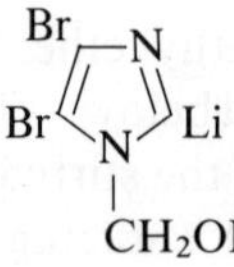 N-(CH_2OEt)-4,5-dibromo-2-lithioimidazole (67)	(32) [e]

3.1. FROM ORGANIC HALIDES

TABLE 3.4—continued

2-Br-pyridine	BuLi (Et₂O)	2-Li-pyridine (94)	(33)
tetrachloro-bipyridine (Cl,Cl,Cl,Cl / Cl,Cl,Cl,Cl)	BuLi (Et₂O)	Cl,Li,Li,Cl / Cl,Cl,Cl,Cl bipyridine (64)	(34)

[a] Labelled with ^{11}C. [b] Yield based on radiochemical analysis. [c] Even less stable carbenoids have been prepared using s-butyllithium in the presence of lithium bromide at $-115°$ (35). [d] Reactions with electrophile, followed by metal–halogen exchange of remaining bromine atoms, used in "one-pot" syntheses of polysubstituted benzenes. [e] Metallation and metal–halogen exchange reactions of imidazoles have been reviewed (36).

For subsequent reactions the co-reagent is usually added at $-60°$ to $-30°$, and the mixture is allowed to warm until reaction occurs. Elimination of lithium chloride, giving tetrachlorobenzyne, is slow below $0°$.

3-Lithiobenzo[b]thiophene

Benzo[b]thiophene is metallated by alkyllithium compounds in the 2-position. For the 3-lithio-compound, therefore, metal–halogen exchange has to be employed. Furthermore 3-lithiobenzo[b]thiophene ring-opens rapidly well below $0°$, so the speed of the metal–halogen exchange at low temperatures is again indispensable. In this procedure the bromobenzothiophene is added to the butyllithium rather than vice versa (19, 20). If the mode of addition is reversed, some 3-bromo-2-lithiobenzo[b]thiophene and benzo[b]thiophene are formed (20).

A solution of n-butyllithium (0.10 mol) in hexane (ca. 45 ml) is added to diethyl ether (250 ml) at ca. $-70°$. The mixture is stirred and kept at ca. $-70°$ while a solution of 3-bromobenzo[b]thiophene (21.3 g, 0.10 mol) in diethyl ether (250 ml) is added during 5–10 min. The mixture is stirred at $-70°$ for a further 30 min. Treatment of the resulting suspension with electrophilic reagents gives the appropriate 3-substituted benzo[b]thiophenes in yields of up to 84%.

References

1. T. Chivers and B. J. Wakefield, *Polychloroaromatic Compounds*, ed. H. Suschitzky, Chapter 3, Plenum, London, 1974.
2. G. C. Nwogoku and H. Hart, *Tetrahedron Lett.* **24**, 5725 (1983).
3. R. Taylor, *Tetrahedron Lett.* 435 (1975).
4. L. V. Sydnes and S. Skare, *Can. J. Chem.* **62**, 2073 (1984).

5. Review: W. E. Parham and C. K. Bradsher, *Acc. Chem. Res.* **15**, 300 (1982).
6. W. E. Parham and L. D. Jones, *J. Org. Chem.* **41**, 2704 (1976).
7. W. E. Parham and Y. A. Sayed, *J. Org. Chem.* **39**, 2055 (1974).
8. W. E. Parham, L. D. Jones and Y. A. Sayed, *J. Org. Chem.* **40**, 2394 (1975).
9. W. E. Parham and L. D. Jones, *J. Org. Chem.* **41**, 1187 (1976).
10. P. E. Peterson, D. J. Nelson and R. Risener, *J. Org. Chem.* **51**, 2381 (1986); see also J. H. Babler and W. E. Banter, *Tetrahedron Lett.* **25**, 4323 (1984).
11. I. Fleming, M. A. Loreto, I. H. M. Wallace and J. P. Michael, *J. Chem. Soc. Perkin Trans. 1* 349 (1986).
12. G. Köbrich and P. Buck, *Chem. Ber.* **103**, 1412 (1970).
13. R. P. Dickinson and B. Iddon, *J. Chem. Soc. (C)* 2592 (1970).
14. D. Seebach and H. Neumann, *Chem. Ber.* **107**, 847 (1974).
15. H. Neumann and D. Seebach, *Chem. Ber.* **111**, 2785 (1978).
16. See Section 3.1.1, Ref. *(14)*.
17. M. D. Rausch, F. E. Tibbetts and H. B. Gordon, *J. Organomet. Chem.* **5**, 493 (1966).
18. N. J. Hales, H. Heaney, J. H. Hollinshead and P. Singh, *Org. Synth.* **59**, 71 (1979).
19. R. P. Dickinson and B. Iddon, *J. Chem. Soc. (C)* 2733 (1968); see also S. Gronowitz, J. Rehnö and J. Sandström, *Acta Chem. Scand.* **24**, 304 (1970).
20. R. P. Dickinson and B. Iddon, *J. Chem. Soc. (C)* 3447 (1971).
21. S. Reiffers, W. Vaalburg, T. Wiegman, H. D. Beerling-Van der Molen, A. M. J. Paans, M. G. Woldring and H. Wynberg, *J. Labelled Cpds. Radiopharmaceuticals* **16**, 56 (1979).
22. J. Villieras, M. Rambaud, R. Tarhouni and B. Kirschleger, *Synthesis* 68 (1981).
23. L. Duhamel, F. Tombret and Y. Mollier, *J. Organomet. Chem.* **280**, 1 (1985).
24. G. Zweifel and W. Lewis, *J. Org. Chem.* **43**, 2739 (1978); see also E.-i. Negishi and T. Takahashi, *J. Am. Chem. Soc.* **108**, 3402 (1986).
25. G. J. Chen and C. Tamborski, *J. Organomet. Chem.* **251**, 149 (1983).
26. P. R. Sharp, D. Astrue and R. R. Schrock, *J. Organomet. Chem.* **182**, 477 (1979).
27. D. W. Meek, G. Dyer and M. O. Workman, *Inorg. Synth.* **16**, 168 (1976).
28. H. A. Brune, R. Hess and G. Schmidtberg, *J. Organomet. Chem.* **303**, 429 (1986).
29. R. Uson and A. Laguna, *Inorg. Synth.* **21**, 72 (1982).
30. Y. Fukuyama, T. Miwa and T. Tokoroyama, *Synthesis* 433 (1974).
31. A. P. Kozikowski and X.-M. Cheng, *J. Org. Chem.* **49**, 3239 (1984); J. M. Muchawski and R. Naef, *Helv. Chim. Acta* **67**, 1168 (1984).
32. B. Iddon and B. L. Lim, *J. Chem. Soc. Perkins Trans. 1* 735 (1983); see also B. Iddon and N. Khan, *Ibid.* 1445 and 1453 (1987).
33. H. Malmberg and M. Nilsson, *Tetrahedron* **42**, 3981 (1986).
34. N. J. Foulger and B. J. Wakefield, *Organomet. Synth.* **3**, 369 (1986).
35. R. Tarhouni, B. Kirschleger and J. Villieras, *J. Organomet. Chem.* **272**, C1 (1984).
36. B. Iddon, *Heterocycles* **23**, 417 (1985).

3.2. PREPARATION BY METALLATION

The replacement of hydrogen by lithium in an organic compound,

RLi + R'H ⟶ RH + R'Li

is perhaps the most versatile method for preparing organolithium compounds. This type of reaction has been well reviewed, so only a brief account

3.2. BY METALLATION

of the general principles involved is given here. For comprehensive reviews of earlier work Refs (*1*) and (*2*) may be consulted; General Refs A and D(i) give much general information; and Refs (*3*) ("Heteroatom-facilitated lithiation"), (*4*) ("Directed metallation of aromatic tertiary amides") and (*5*) ("Metallation and electrophilic substitution of amine derivatives adjacent to nitrogen") cover important selected areas.

The simplest lithiations are those of relatively strong hydrocarbon acids ($pK_a \leqslant ca.$ 33) such as 1-alkynes and triarylmethanes, which may be regarded as straightforward acid–base reactions. However, many compounds whose acidity would be expected to be much lower are readily lithiated. This occurs notably when activation by an α- or β-heteroatom increases the kinetic and/or thermodynamic acidity of a particular hydrogen atom (*3*). This type of activation is particularly useful for the introduction of a substituent *ortho* to an existing functional group or on an *ortho*-methyl group, via the organolithium intermediate:

A selection of groups X that promote *ortho*-lithiation in this way is shown in Table 3.5.

As indicated in Table 3.5, some of these groups also function as protecting groups. It is not possible to place the directing groups in an invariable order of effectiveness, since the relative strengths of the groups may change in different solvents or in the presence of additives. A fairly generally applicable order is that given in General Ref. D(i): $-SO_2NR_2 > -SO_2Ar > CONR_2 >$ oxazolinyl $> -CONHR, -CSNHR, CH_2NR_2 > -OR > -NHAr > -SR >$ $-NR_2 > CR_2O^-$. A potentially useful method for modifying the "normal" order is complexation with $Cr(CO)_3$ (*17*).

The directing effect of a substituent can sometimes override the effect of a ring heteroatom. For example, a thiophene may be lithiated at a β-position, rather than in the usual α-position (*cf.* the syntheses of 2-lithio- and 2,5-dilithiothiophene described below) (*18*):

TABLE 3.5

Some Groups Promoting *ortho*-Lithiation

Directing group	Notes	Ref.
$-NR_2$		(a)
$-N(R)CO_2Li$	Protected $-NHR$ group	(6)
$-N(Li)COR$	From $-NHCOR$; bulky R group required. Protected $-NH_2$ group	(7)
$-CH_2NR_2$		(a)
$-CON(Li)R$	From $-CONHR$. Protected $-CO_2H$ group	(a, 5, 8)
$-CONR_2$	Protected $-CO_2H$ group, though hydrolysis difficult	(a, 4, 5)
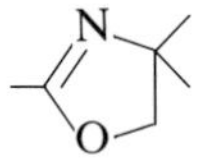 oxazoline group	Protected $-CO_2H$ group	(a)
$-OLi$		(9)
$-OMe$		(a)
$-OCMe_3$	Protected $-OH$ group	(10)
$-OCH_2OMe$	Protected $-OH$ group	(11)
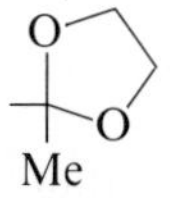 dioxolane with Me	Protected $-COMe$ group	(12)
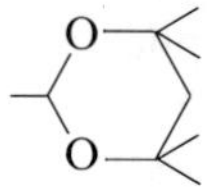 dioxane	Protected $-CHO$ group	(13)
$-SR$	Care needed to minimize α-metallation of R group	(14)
$-SO_2R$		(a)
$-SO_2NR_2$		(a)
$-SO_2OR$		(15)
$-F, Cl, Br, I$	Reactions must be carried out at low temperature to avoid aryne formation. For Br and I, use of $LiNR_2$ avoids metal–halogen exchange	(a, 16)

a Many references. See General Refs and Refs (2) and (3) for examples.

3.2. BY METALLATION

Directed metallation can also take precedence over competing reactions, such as addition to π-deficient heterocycles (*19*); for example (*7*)

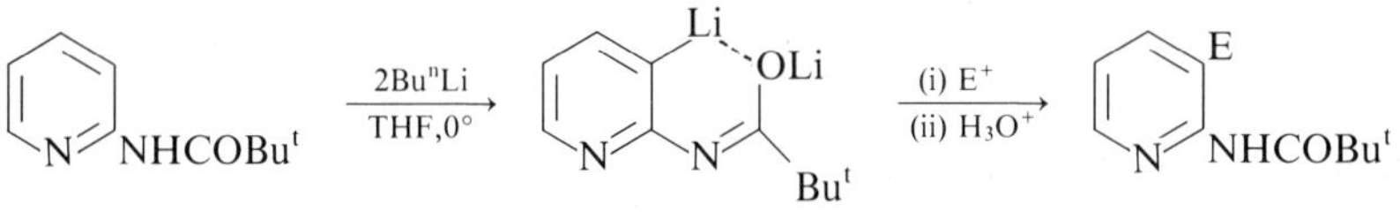

[Scheme: 2-pyridyl-NHCOBu^t → (2Bu^{n}Li, THF, 0°) → lithiated intermediate with OLi and Bu^t → (i) E$^+$ (ii) H$_3$O$^+$ → 3-E-2-pyridyl-NHCOBu^t]

In many cases lithiation takes place readily by means of simple alkyllithium reagents including the most common ones commercially available, *viz.* butyllithium in hydrocarbon solvents. However, n-butyllithium is a strong nucleophile and thus usually incompatible with, for example, carbonyl groups. When such functional groups are present, reagents with high basicity but relatively low nucleophilicity must be employed. In some cases the increased steric hindrance of nucleophilic addition by s- and t-butyllithium enables them to be employed (for example in the *o*-metallation of *N,N*-dialkylbenzamides (*4*)). More commonly, hindered lithium dialkylamides are used; although their thermodynamic basicity is lower than that of butyl-lithium they show very high kinetic basicity. The most widely used is lithium diisopropylamide (LDA), but even this amide can act as a nucleophile in some cases (*20*). An even more hindered alternative is lithium 2,2,6,6-tetra-methylpiperidide [1] (LiTMP) (*21*). Lithium bis(trimethylsilyl)amide (LTSA) (*22*), (Me$_3$Si)$_2$NLi, though less basic, has also been found useful.

[Structure: LiN in tetramethylpiperidide ring]

[1]

LDA cleaves ethers, particularly THF, quite rapidly, but it is easy to make it *in situ* from diisopropylamine and butyllithium as in the aldol reaction described below (see also Ref. (*22a*)). It is now also commercially available as a solid and as a stable solution in a hydrocarbon–THF mixture (*23*). An alternative method of synthesis, not requiring an organolithium compound, is the reaction of diisopropylamine with lithium metal in the presence of an electron carrier such as styrene (*24*). LTSA cleaves THF less rapidly than does LDA, and is commercially available in that solvent. LiTMP is sold as a suspension in diethyl ether. Other hindered lithium amides are described in Ref. (*24a*).

The solvent used for metallation reactions may be of crucial importance. In general, metallation is promoted by electron-donor solvents, which by coordinating with lithium both reduce the degree of association of the reagent and increase the polarization of the lithium–carbon (or lithium–nitrogen) bond. Thus metallations occur faster in ethers than in hydrocarbons, and faster in THF than in diethyl ether. Reactivity is further enhanced by the addition of very strong donors such as hexametapol and diamines such as TMEDA and DABCO. Some disadvantages of such solvents and additives are noted in Section 2.1.1.

Some representative procedures for metallation are described below, and others are listed in Table 3.6, which may be regarded as supplementary to the tables of older examples given in the general references.

One further technique that should be referred to, although it gives organopotassium compounds (or mixed organopotassium–organolithium aggregates) rather than organolithium compounds, is the use of metallating reagents prepared from potassium t-alkoxides and organolithium compounds (25). The metallation of propene by this method is described below, and even ethylene has been metallated in this way (26). A combination ("KDA") of LDA and potassium t-butoxide is also a very powerful metallating reagent (27). If an organolithium compound rather than an organopotassium compound is desired, it may be obtained by adding lithium bromide to the solution or suspension of the latter.

2-Lithio-1,3-dithiane

2-Lithiated 1,3-dithianes are among the most versatile acyl-anion equivalents (28). In the *Organic Syntheses* procedure (29) on which the following description is based, sequential alkylations of such intermediates gave a spirocyclobutane and thence cyclobutanone:

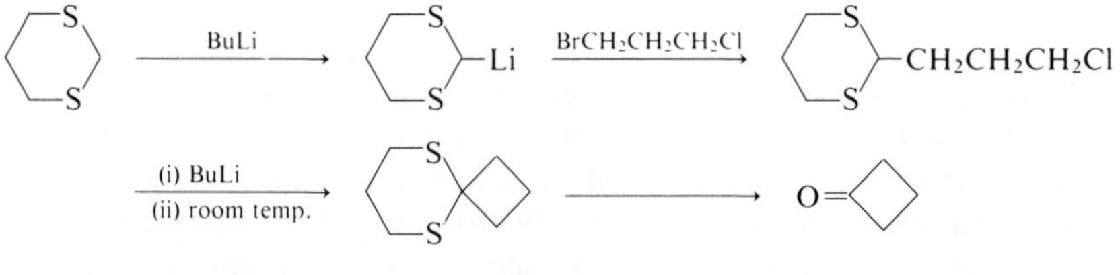

Dry THF (1.25 l) and 1,3-dithiane (50 g, 0.417 mol) are added to a 2 l round-bottomed flask fitted with a septum and furnished with an efficient magnetic stirrer and an inert atmosphere. The solution is stirred and cooled (bath temperature −20°). n-Butyllithium (*ca.* 2 M in hexane, 0.430 mol) is added by syringe, and the bath temperature is maintained at −10° to −20°.

In the *Organic Syntheses* experiment the bath temperature was lowered to −75° prior to the reaction with 1-bromo-3-chloropropane (see p. 110).

3.2. BY METALLATION

A mixed aldol reaction with cyclopentanone enolate

Because of its low nucleophilicity, LDA is the reagent of choice for generating lithium enolates from ketones; the reaction occurs at temperatures sufficiently low for self-aldol reaction to be slow compared with mixed aldol reactions with aldehydes. The lithium enolate may also be used *in situ* to prepare other enolates, such as those of boron, tin and zirconium (*30*). The following example (*30*) is a key step in a synthesis of prostacyclin (*31*):

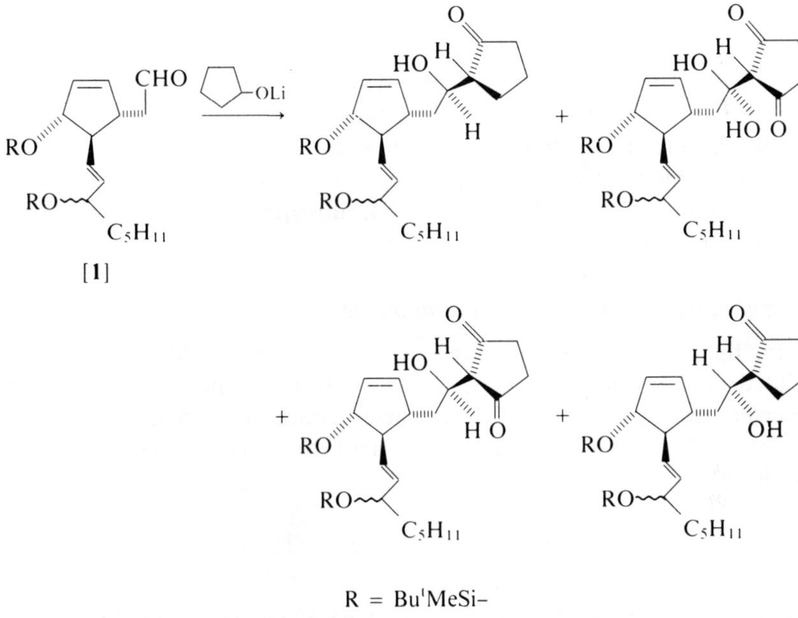

$R = Bu^tMeSi-$

A 50 ml three-necked flask is fitted with a septum, a low-temperature thermometer, a magnetic stirrer and means for maintaining an inert atmosphere. The solvents and reagents are added by syringe. A solution of diisopropylamine (39.5 mg, 0.39 mmol) in diethyl ether (8 ml) is added, cooled to 0°, and kept below 3° as n-butyllithium (*ca.* 1.5 M in hexane, 0.39 mmol) is added dropwise. The solution is stirred at 0° for 10 min, then cooled in a carbon dioxide–acetone bath. Cyclopentanone (32 mg, 0.39 mmol) is added dropwise during 2 min, and the mixture is stirred at −78° for 30 min. A solution of the cyclopentenylacetaldehyde [1] (185 mg, 0.37 mmol) in diethyl ether (3 ml) is added, and the mixture is stirred at −78° for 10 min. Saturated aqueous ammonium chloride (20 ml) is added, and the products are isolated as described in Ref. (*30*).

Benzyllithium

Reactions of benzyl halides with lithium metal or with alkyllithium compounds give at best very poor yields of benzyllithium. While other routes to benzyllithium have been developed (32), metallation of toluene is a particularly attractive method. This reaction occurs only to a small extent with n-butyllithium even in THF, but is rapid and almost quantitative in the presence of TMEDA or DABCO (33, 34):

$$PhCH_3 \xrightarrow[TMEDA]{BuLi} PhCH_2Li.TMEDA$$

A solution of TMEDA (1.16 g, 0.01 mol) or DABCO (1.12 g, 0.01 mol) in dry toluene (30 ml) is placed in a flask fitted with a septum and a stirrer and furnished with an inert atmosphere. n-Butyllithium (ca. 1.5 M in hexane, 0.01 mol) is added by syringe, and the mixture is stirred at 80° for 30 min. The benzyllithium–DABCO complex separates as bright-yellow needles.

In the experiment described in Ref. (33), subsequent reaction with benzophenone gave the carbinol in 85% yield.

2-Lithiothiophene and 2,5-dilithiothiophene

Furan, thiophene and N-protected pyrroles are readily lithiated and dilithiated in the α-positions (35). Under controlled conditions (which are somewhat different for the different heterocycles) it is possible to obtain excellent yields of the mono- or dilithio derivatives, as described below for thiophene (35, 36).

(a) TMEDA (1.55 g, 13 mmol)[a] is placed in a 50 ml two-necked flask fitted with a condenser, a gas inlet and a magnetic stirrer. The apparatus is thoroughly flushed with nitrogen, the gas inlet is replaced by a rubber septum, and a positive pressure of nitrogen is maintained in the apparatus by means of a balloon. A solution of n-butyllithium (13 mmol) in diethyl ether (ca. 9 ml) is added by syringe, followed by thiophene (1.06 g, 13 mmol). After being stirred at 25° for 30 min, the solution contains 2-lithiothiophene (11.7 mmol) with a little 2,5-dilithiothiophene (0.9 mmol), as determined by reaction with carbon dioxide followed by esterification with diazomethane.

(b) A similar apparatus to that described in (a) is used. The flask initially contains TMEDA (3.88 g, 33 mmol) and a solution of n-butyllithium (33 mmol) in hexane (ca. 23 ml) and thiophene (1.06 g, 13 mmol) are added. The mixture is boiled under reflux for 30 min, and then contains 2,5-dilithiothiophene (12.6 mmol).

3.2. BY METALLATION   39

a In THF satisfactory yields of the monolithio-derivative may be obtained in the absence of TMEDA.

2-Lithio-N,N-diethylbenzamide (37)

Ortho-lithiation of many aromatic compounds, where the *ortho*-directing group is unreactive towards the metallating reagent, is usually straightforward. However, lithiation *ortho* to reactive functional groups, which can subsequently be used for further reactions, is particularly useful (*4, 38*), but requires special care:

$$\text{PhCONEt}_2 \xrightarrow{\text{Bu}^s\text{Li, TMEDA}} \text{2-Li-C}_6\text{H}_4\text{CONEt}_2$$

TMEDA (1.92 g) is dissolved in dry THF (50 ml) in a 250 ml flask fitted with a stirrer, a septum and a pressure-equalizing dropping funnel, and maintained with an atmosphere of nitrogen. A solution of N,N-diethylbenzamide (2.66 g) in dry THF (40 ml) is placed in the dropping funnel. The solution in the flask is cooled to $-78°$ and stirred as s-butyllithium (11.0 ml of 1.5 M solution in cyclohexane) is added. The resulting suspension is stirred at $-78°$ as the solution of N,N-diethylbenzamide is added dropwise. After the addition is complete, stirring is continued for 20 min at $-78°$.

When a solution thus prepared was hydrolysed with deuterium oxide, 2-deuterio-N,N-diethylbenzamide, showing 95% deuterium incorporation, was isolated in 88% yield (*37*).

Metallation of propene by butyllithium–potassium t-butoxide reagent (39)*

In the equations for this procedure the reagent is formulated as butylpotassium, and the product as a localized allyl derivative, purely for convenience:

$$\text{Bu}^n\text{Li} + \text{Bu}^t\text{OK} \longrightarrow \text{Bu}^n\text{K} + \text{Bu}^t\text{OLi}$$

$$\text{CH}_3\text{CH}=\text{CH}_2 + \text{Bu}^n\text{K} \longrightarrow \text{KCH}_2\text{CH}=\text{CH}_2 + \text{Bu}^n\text{H}$$

$$\text{KCH}_2\text{CH}=\text{CH}_2 + \text{LiBr} \longrightarrow \text{LiCH}_2\text{CH}=\text{CH}_2 + \text{KBr}$$

n-Butyllithium (*ca.* 1.5 M in hexane, 0.10 mol) is diluted with THF and cooled to $-95°$. The solution is stirred and the temperature kept below $-80°$ as liquid propene (8.42 g, 0.20 mol) is added, followed by a solution of

* The lithiation of propene by butyllithium–TMEDA has been described by Akiyama and Hooz (*40*).

TABLE 3.6

Preparation of Organolithium Compounds by Metallation

Substrate	Metallating reagent	Solvent (additive)	Organolithium product (yield %[a])	Ref.
Bu^{n}C≡CH	Bu^{n}Li	Pentane	CH$_3$CH$_2$CH$_2$CH(Li)C≡CLi[b] (64–68)	(41)
Ph$_3$CH	Bu^{n}Li	THF–Et$_2$O	Ph$_3$CLi (75–85)	(43)
CH$_3$\\C=CH$_2$ / HOCH$_2$	Bu^{n}Li	TMEDA	LiCH$_2$\\C=CH$_2$ / LiOCH$_2$ (52)	(44)
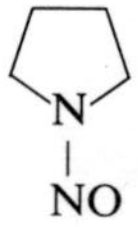 ⟨pyrrolidine-N–NO⟩	LDA	THF	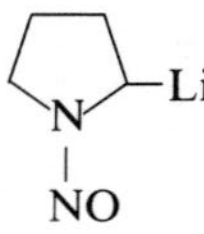 ⟨2-Li-pyrrolidine-N–NO⟩ (68–70)	(45)
PhCH$_2$SMe	Bu^{n}Li	Hexane (TMEDA)	PhCH(Li)SMe (75)	(46)
(PhS)$_2$CH$_2$	Bu^{n}Li	THF	(PhS)$_2$CHLi (48–52)	(47)
MeOCH=CH$_2$	Bu^{t}Li	THF	MeOC(Li)=CH$_2$ (74)	(3), cf. (48)
PhSCH=CH$_2$	Bu^{n}Li or LDA	THF (TMEDA) or THF (hexametapol)	PhSC(Li)=CH$_2$ (76 or 84)	(3), cf. (49)
Me$_2$NCHO	LDA	Et$_2$O–THF	Me$_2$NC(Li)=O (46–52)	(3), cf. (50)

TABLE 3.6—continued

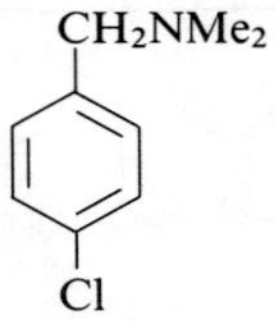

Substrate	Reagent	Solvent	Product	Ref.
4-Cl-C₆H₄-CH₂NMe₂	Bu^{n}Li	Et₂O	2-Li-4-Cl-C₆H₃-CH₂NMe₂ (79)	(3)
4-MeO-C₆H₄-(4,4-dimethyloxazoline)	Bu^{n}Li	Et₂O	2-Li-4-MeO-C₆H₃-(oxazoline) (89)	(3), cf. (51)
PhNMe₂	Bu^{n}Li	Hexane	2-Li-C₆H₄-NMe₂ (66–67)	(52)
2-Cl-C₆H₄-OCH₂OMe	Bu^{n}Li	Hexane (TMEDA)	2-Cl-6-Li-C₆H₃-OCH₂OMe (65)	(3), cf. (53)

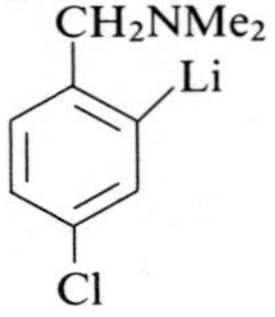

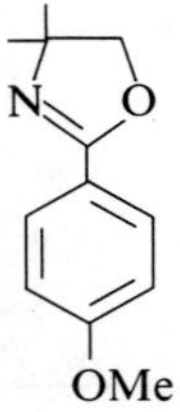

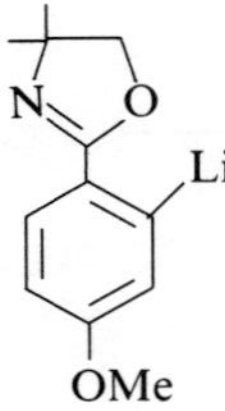

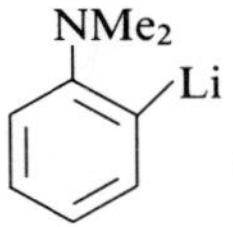

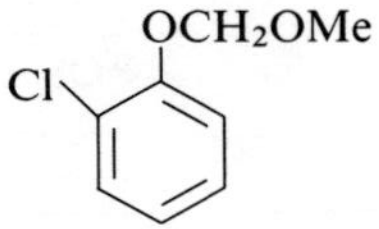

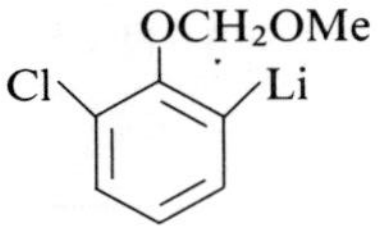

TABLE 3.6—continued

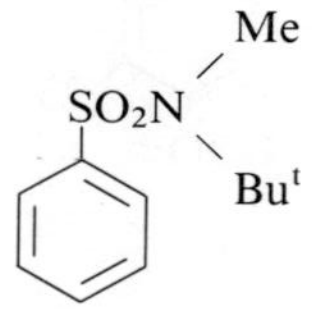

Ph-SO₂N(Me)(Bu^t)	Bu^nLi	Et₂O	2-Li-C₆H₄-SO₂N(Me)(Bu^t) (81) (3)
 furan	Bu^nLi	THF	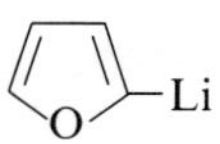 2-Li-furan (99.8) (3)
 N-methylpyrrole	Bu^nLi	Hexane (TMEDA)	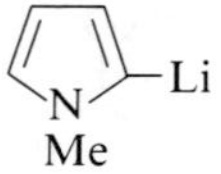 2-Li-N-methylpyrrole (91) (54)
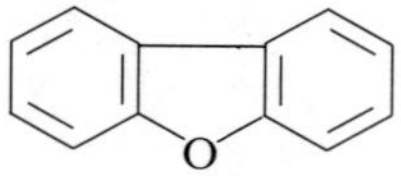 dibenzofuran	Bu^nLi	Et₂O	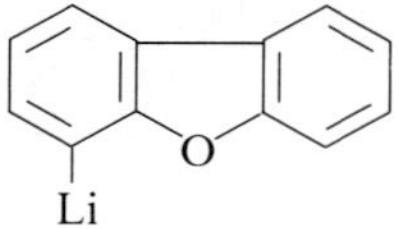 4-Li-dibenzofuran (50–60) (55)
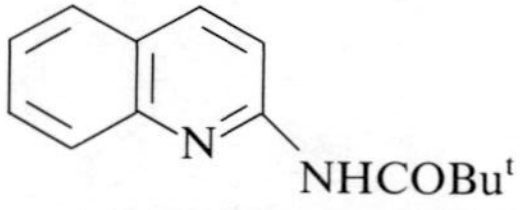 2-(NHCOBu^t)-quinoline	Bu^nLi	Et₂O	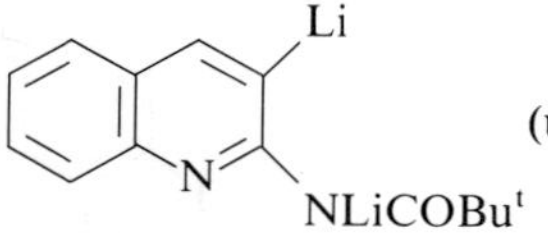 3-Li-2-(NLiCOBu^t)-quinoline (up to 95) (56)

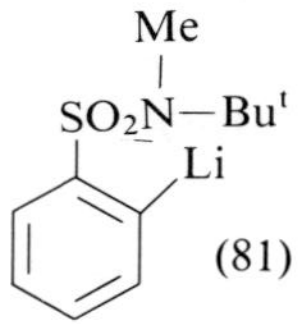

^a Yields are those of final products of the experiments described in the references cited. ^b The dilithio-compound reacts selectively at the propargylic position (42).

3.2. BY METALLATION

potassium t-butoxide (11.3 g, 0.10 mol) in THF (30 ml). The temperature is allowed to rise to $-20°$ during 30 min.

The yellow suspension thus obtained could be used for further reactions, but in the example described (involving reaction with carbon disulphide) better results were obtained by converting the "allylpotassium" to allyllithium by the addition of lithium bromide. Anhydrous lithium bromide is prepared by heating the commercial material to 150° under 10–15 Torr during 15 min.

The suspension is stirred as a solution of anhydrous lithium bromide (10.5 g, 0.12 mol) in THF (40 ml) is added; the colour fades to a very pale yellow. In the experiment described, subsequent reaction with carbon disulphide followed by iodomethane gave 1,1-bis(methylthio)-1,3-butadiene (80%) (39).

References

1. See Section 3.1.1, Ref. (7).
2. J. M. Mallan and R. L. Bebb, *Chem. Rev.* **69**, 693 (1969).
3. H. W. Gschwend and H. R. Rodriguez, *Org. React.* **26**, 1 (1976).
4. P. Beak and V. Snieckus, *Acc. Chem. Res.* **15**, 306 (1982).
5. P. Beak, W. J. Zajdel and D. B. Reitz, *Chem. Rev.* **84**, 471 (1984).
6. A. R. Katritzky, W.-Q. Fan and K. Akutagawa, *Tetrahedron* **42**, 4027 (1986).
7. J. A. Turner, *J. Org. Chem.* **48**, 3401 (1983).
8. W. H. Puterbaugh and C. R. Hauser, *J. Org. Chem.* **29**, 853 (1964).
9. G. H. Posner and K. A. Canella, *J. Am. Chem. Soc.* **107**, 2571 (1985).
10. D. A. Shirley, J. R. Johnson and J. D. Hendrix, *J. Organomet. Chem.* **11**, 217 (1968).
11. M. R. Winkle and R. C. Ronald, *J. Org. Chem.* **47**, 2101 (1982).
12. N. S. Narasimhan, A. C. Ranade, H. R. Deshpande, U. B. Gokhale and G. Jayalakshmi, *Synth. Commun.* **14**, 373 (1984).
13. A. L. Campbell and I. K. Khanna, *Tetrahedron Lett.* **27**, 3963 (1986).
14. L. Horner, A. J. Lawson and G. Simons, *Phosphorus Sulfur* **12**, 353 (1982).
15. J. N. Bonfiglio, *J. Org. Chem.* **51**, 2833 (1986).
16. H. Gilman and T. S. Soddy, *J. Org. Chem.* **22**, 1715 (1957).
17. J. P. Gilday and D. A. Widdowson, *J. Chem. Soc. Chem. Commun.* 1235 (1986); P. M. Treichel and R. V. Kirss, *Organometallics* **6**, 249 (1987).
18. A. J. Carpenter and D. J. Chadwick, *J. Chem. Soc. Perkin Trans. 1* 173 (1985).
19. F. Marsais and G. Queguiner, *Tetrahedron* **39**, 2009 (1983).
20. J. L. Herrmann, G. R. Kieczykowski and R. H. Schlessinger, *Tetrahedron Lett.* 2433 (1973).
21. C. M. Dougherty and R. A. Olofson, *Org. Synth.* **58**, 37 (1978).
22. E. A. Amonoo-Neizer, R. A. Shaw, D. A. Skovlin and B. C. Smith, *Inorg. Synth* **8**, 19 (1966); D. C. Bradley and R. G. Copperthwaite, *Ibid.* **18**, 112 (1978).
22a. J. Einhorn and J. L. Luche, *J. Org. Chem.* **52**, 4124 (1987).
23. Lithium Corporation Technical Bulletin; *cf.* H. O. House, W. V. Phillips, T. S. B. Sayer and C.-C. Yau, *J. Org. Chem.* **43**, 700 (1978).
24. M. T. Reetz and W. F. Maier, *Justus Liebigs Ann. Chem.* 1471 (1980); see also F. Gaudemar-Bardone and M. Gaudemar, *Synthesis* 469 (1979).
24a. I. E. Kopka, Z. A. Fataftah and M. W. Rathke, *J. Org. Chem.* **52**, 448 (1987).

25. L. Lochmann, J. Popisil and D. Lim, *Tetrahedron Lett.* 257 (1966); M. Schlosser and S. Strunk, *Tetrahedron Lett.* **25**, 741 (1984); R. P. W. Bauer, B. Brix, C. Schade and P. v. R. Schleyer, *J. Organomet. Chem.* **306**, C1 (1986); H. D. Verkruijsse, L. Brandsma and P. v. R. Schleyer, *J. Organomet. Chem.* **332**, 99 (1987).
26. L. Brandsma, H. D. Verkruijsse, C. Schade and P. v. R. Schleyer, *J. Chem. Soc. Chem. Commun.* 260 (1986).
27. S. Raucher and G. A. Koolpe, *J. Org. Chem.* **43**, 3794 (1978).
28. D. Seebach, *Synthesis* 17 (1969).
29. D. Seebach and A. K. Beck, *Org. Synth.* **51**, 76 (1971).
30. A. D. Baxter, S. M. Roberts, B. J. Wakefield, G. T. Woolley and R. F. Newton, *J. Chem. Soc. Perkin Trans. 1* 1809 (1983).
31. A. D. Baxter, S. M. Roberts, B. J. Wakefield, G. T. Woolley and R. F. Newton, *J. Chem. Soc. Perkin Trans. 1* 675 (1984).
32. J. J. Eisch, *Organomet. Synth.* **2**, 95 (1981).
33. G. G. Eberhardt and W. A. Butte, *J. Org. Chem.* **29**, 2928 (1964).
34. A. W. Langer, *Trans. NY Acad. Sci.* **27**, 741 (1965).
35. (a) D. J. Chadwick and C. Willbe, *J. Chem. Soc. Perkin Trans. 1* 887 (1977); (b) E. Jones and I. M. Moodie, *Org. Synth.* **50**, 104 (1970).
36. C. Willbe, Part II Thesis, University of Liverpool; D. J. Chadwick, Personal communication.
37. P. Beak and R. A. Brown, *J. Org. Chem.* **47**, 34 (1982).
38. N. S. Narasimhan and R. S. Mali, *Top. Curr. Chem.* **138**, 63 (1987).
39. Y. A. Heus-Kloos, R. L. P. de Jong, H. D. Verkruijsse, L. Brandsma and S. Julia, *Synthesis* 958 (1985).
40. S. Akiyama and J. Hooz, *Tetrahedron Lett.* 4115 (1973).
41. A. J. Quillinan and F. Scheinmann, *Org. Synth.* **58**, 1 (1978).
42. A. J. G. Sagar and F. Scheinmann, *Synthesis* 321 (1976).
43. J. J. Eisch, *Organomet. Synth.* **2**, 98 (1981).
44. B. M. Trost, D. M. T. Chan and T. N. Nanninga, *Org. Synth.* **62**, 58 (1984).
45. D. Enders, R. Pieter, B. Renger and D. Seebach, *Org. Synth.* **58**, 113 (1978).
46. S. Cabiddu, C. Floris, S. Melis and F. Sotgiu, *Phosphorus Sulfur* **19**, 61 (1984).
47. T. Cohen, R. J. Ruffner, D. W. Shull, E. R. Fogel and J. R. Falck, *Org. Synth.* **59**, 202 (1980).
48. J. E. Baldwin, O. W. Lever and N. R. Tzodikov, *J. Org. Chem.* **41**, 2313 (1976).
49. R. C. Cookson and P. J. Parsons, *J. Chem. Soc. Chem. Commun.* 990 (1976).
50. B. Banhidai and U. Schöllkopf, *Angew. Chem. Int. Ed. Engl.* **12**, 836 (1973).
51. H. W. Gschwend and A. Hamdan, *J. Org. Chem.* **40**, 2008 (1975).
52. S.-I. Murahashi, T. Naota and Y. Tanigawa, *Org. Synth.* **62**, 39 (1984); see also J. V. Hay and T. M. Harris, *Org. Synth.* **53**, 56 (1973) for the 4-methyl analogue.
53. H. Christensen, *Synth. Commun.* **5**, 65 (1975).
54. D. J. Chadwick and I. A. Cliffe, *J. Chem. Soc. Perkin Trans. 1* 2845 (1979); see also J. M. Brittain, R. A. Jones, J. S. Arques and T. A. Saliente, *Synth. Commun.* **12**, 231 (1975).
55. D. S. Kemp and N. G. Galakatos, *J. Org. Chem.* **51**, 1821 (1986).
56. J. M. Jacquelin, F. Marsais, A. Godard and G. Queguiner, *Synthesis* 670 (1986).

3.3. PREPARATION FROM OTHER ORGANOMETALLIC COMPOUNDS

In comparison with the general methods described in the preceding sections, syntheses of organolithium compounds from other organometallic

3.3. FROM OTHER ORGANOMETALLIC COMPOUNDS

compounds are much less widely useful. Nevertheless, there are circumstances in which the generally more cumbersome procedures involved are justified by particular requirements. For example, the reaction of lithium metal with a dialkylmercury compound* is a means of obtaining a completely halide-free organolithium compound. Transmetallation between an organolithium compound and an organic derivative of another metal (or metalloid) is known for a number of elements (see General Ref. A), but is most commonly applied to trialkyltin compounds and to selenoacetals. Some examples are listed in Table 3.7.

The transformation of organopotassium compounds into organolithium compounds by reaction with lithium bromide has been noted in Section 3.2.

1,2-Dilithiobenzene

This example is a preparation of an otherwise inaccessible, but very useful, reagent by the organomercury route (10). The structure of the starting material "o-phenylenemercury" has been controversial, but it is probably largely a trimeric macrocycle (11):

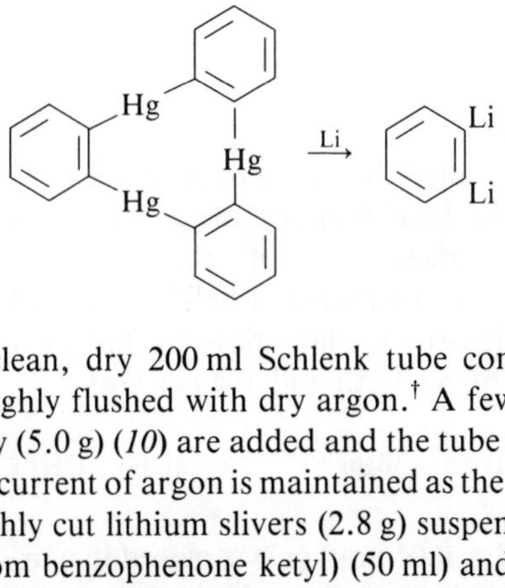

A scrupulously clean, dry 200 ml Schlenk tube containing a magnetic stirrer bar is thoroughly flushed with dry argon.† A few glass splinters and o-phenylenemercury (5.0 g) (10) are added and the tube is evacuated and refilled with argon. A current of argon is maintained as the following are added successively: (i) freshly cut lithium slivers (2.8 g) suspended in diethyl ether (freshly distilled from benzophenone ketyl) (50 ml) and (ii) more dry ether (60 ml). The Schlenk tube is sealed under argon and the contents are stirred for 4 days. The contents of the tube develop a brown colour, and eventually the solution becomes a deep red. Suspended mercury (black) and lithium are also present; if any suspended crystals of o-phenylenemercury remain, stirring should be continued until reaction is complete. The yield of

---

*This type of reaction is of historical importance, as the first organolithium compounds were prepared in this way (1).
† In the original preparation nitrogen was used, but argon is preferable if available.

3. PREPARATION OF ORGANOLITHIUM COMPOUNDS

TABLE 3.7

Preparation of Organolithium Compounds by Transmetallation

Starting material	Lithium reagent (solvent)	Product (yield %)[a]	Ref.
Ph_2Hg	Li (benzene + Et_2O, 10:1)	PhLi (64)	(2) (based on (1))
$PhN(Li)CH_2CH_2HgBr$	Li (THF)	$PhN(Li)CH_2CH_2Li$ (90)	(3)
$CH_2=CHCH_2SnPh_3$	PhLi (Et_2O)	$CH_2=CHCH_2Li$ (65–75)	(4)
$Bu_3^nSnCH_2NMe_2$	Bu^nLi (THF)	$LiCH_2NMe_2$ (95)	(5)
$CH_2=C(OMe)SnMe_3$	Bu^nLi (hexane)	$CH_2=C(OMe)Li$ (97)	(6)
$Bu_3^nSnCH_2OH$	Bu^nLi (pentane)	$LiCH_2OLi^b$ (65)	(7)
$(PhSe)_2CH_2$	Bu^nLi (THF)	$PhSeCH_2Li$ (95)	(8)
$(MeSe)_2CMe_2$	Bu^nLi (THF)	$MeSeC(Li)Me_2$ (88)	(9)

[a] Yield of product of subsequent reaction. [b] Possibly better formulated as $\left[Bu_4Sn\diagup\!\!\!\!\!{}^{O}\diagdown\right]^{2\ominus} 2Li^{\oplus}$.

1,2-dilithiobenzene in solution is ca. 90%. To be used for subsequent reactions the solution is decanted through a glass-wool plug into the reaction vessel.

Vinyllithium

Although solutions of vinyllithium can be prepared by the reaction of lithium with vinyl chloride in THF (12), by metal–halogen exchange with vinyl bromide (13) and via metallation of ethene (14), transmetallation with tetravinyltin enables it to be prepared halide-free, and in solid form. This procedure is a smaller-scale adaptation by Fraenkel et al. (15) of one originally published by Seyferth and Weiner (16).

$$(CH_2=CH)_4Sn + 4Bu^nLi \longrightarrow 4CH_2=CHLi + Bu_4^nSn$$

Tetravinyltin (2.25 g, 0.01 mol) (17) is placed in a dry 100 ml flask, fitted with a stirrer and septum and flushed with argon. A solution of n-butyllithium in pentane (ca. 1.6 M, 0.038 mol) is added via the septum and the mixture is stirred for 3 h; after 30 min the mixture has started to become milky. The suspension is allowed to settle, and the supernatant liquor is removed by means of a syringe. The remaining solid is washed with five 20 ml portions of dry pentane. The remaining volatile material is removed by evacuating the flask, and the residue is dissolved in an appropriate solvent.

In the experiment described (15) the yield of solid vinyllithium was 56%, which was dissolved in diethyl ether to give an approximately 1.5 M solution.

References

1. W. Schlenk and J. Holtz, *Ber. Dtsch. Chem. Ges.* **50**, 262 (1917).
2. D. Thoennes and E. Weiss, *Chem. Ber.* **111**, 3157 (1978).
3. J. Barlengua, F. J. Fanañas and M. Yus, *J. Org. Chem.* **44**, 4798 (1979); J. Barlengua, F. J. Fanañas and M. Yus and G. Asensio, *Tetrahedron Lett.* 2015 (1978).
4. J. J. Eisch, *Organomet. Synth.* **2**, 92 (1981).
5. J.-P. Quintard, B. Elissondo and B. Jousseaume, *Synthesis* 495 (1984); *cf.* D. J. Peterson, *J. Organomet. Chem.* **21**, P63 (1970).
6. J. A. Soderquist and G. J.-H. Hsu, *Organometallics* **1**, 830 (1982).
7. N. Meyer and D. Seebach, *Chem. Ber.* **113**, 1290 (1980).
8. D. Seebach and N. Peleties, *Chem. Ber.* **105**, 511 (1972).
9. D. Van Ende, W. Dumont and A. Krief, *Angew. Chem. Int. Ed. Engl.* **14**, 700 (1975).
10. H. J. S. Winkler and G. Wittig, *J. Org. Chem.* **28**, 1733 (1963); *cf.* G. Wittig and F. Bickelhaupt, *Chem. Ber.* **91**, 833 (1958).
11. D. S. Brown, A. G. Massey and D. A. Wickens, *Acta Cryst.* **B34**, 1695 (1978).
12. See Section 3.1.1, Ref. (*21*).
13. See Section 3.1.3, Ref. (*15*).
14. See Section 3.2, Ref. (*26*).
15. G. Fraenkel, H. Hsu and B. M. Su, *Lithium: Current Applications in Science, Technology, and Medicine*, ed. R. O. Bach, p. 273, Wiley-Interscience, New York, 1985.
16. D. Seyferth and M. A. Weiner, *J. Am. Chem. Soc.* **83**, 3585 (1961).
17. H. G. Kuivila and T. McComb, *Organomet. Synth.* **3**, 597 (1986).

3.4. PREPARATION FROM ETHERS AND THIOETHERS

Although other examples are known, for practical purposes the synthesis of organolithium compounds by the cleavage of ethers is limited to reactions of lithium metal with allyl and benzyl ethers (*1, 2*). (For a review see Ref. 2a.) The synthesis of allyllithium by this method is described below. The synthesis of allyllithium derivatives by an analogous cleavage of allyl 2,4,6-trimethylbenzoates has also been reported (*3*).

The cleavage of thioethers by lithium or by lithium salts of radical anions is a much more versatile method for preparing organolithium compounds. Some examples are listed in Table 3.8. The best reagent for a particular case appears to vary. Cohen *et al.* have recently introduced lithium 1-dimethylaminonaphthalenide (LDMAN); not only does this reagent give good yields of organolithium compounds, but the co-produced 1-dimethylaminonaphthalene is readily separated from the final product by washing with dilute acid, as in the preparation of 4-t-butyl-1-phenylthiocyclohexyllithium described below.

Allyllithium (1, 9)

$$CH_2=CHCH_2OPh + 2Li \longrightarrow CH_2=CHCH_2Li + LiOPh$$

Freshly cut finely divided lithium (4.2 g, 0.6 mol) is suspended in dry THF

TABLE 3.8

Preparation of Organolithium Compounds from Thioethers

Thioether	Reagent (solvent)	Organolithium product (yield %)[a]	Ref.
Et | BuCHCH₂SPh	LN (THF)	Et | BuCHCH₂Li (84)	(4)
(PhSCH₂CH₂CH₂)₂O	Li (THF)	(LiCH₂CH₂CH₂)₂O (90)	(4)
Bu^{t}SPh	Li (THF)	Bu^{t}Li (78)	(5)
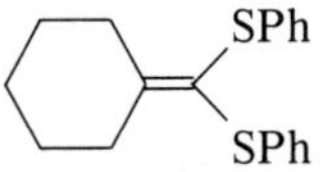 ⌬=C(SPh)(SPh)	LN (THF)	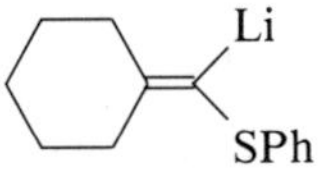 ⌬=C(Li)(SPh) (91)	(6)
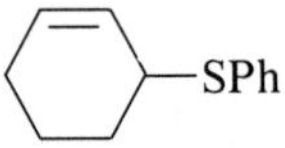 ⌬—SPh	LN (THF)	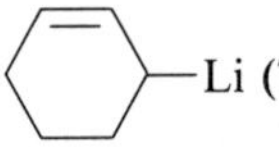 ⌬—Li (72)	(7)
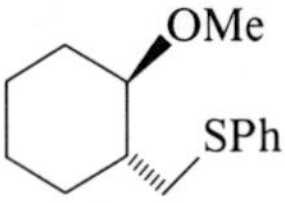 ⌬(OMe)(SPh)	LiDBB (THF)	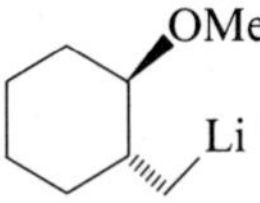 ⌬(OMe)(Li) (81)[b]	(8)

[a] Yield of product of subsequent reaction. [b] This type of reaction was used to prepare tri-, and even possibly hexa-lithio compounds.

in a 500 ml three-necked flask equipped with a stirrer and a pressure-equalizing dropping funnel, and furnished with an atmosphere of argon. The suspension is stirred and cooled in a bath at −15° while a solution of allyl phenyl ether (6.7 g, 50 mmol) in dry diethyl ether (25 ml) is added dropwise during 45 min. (If a pale green or blue colour is not observed after a small proportion of the allyl ether has been added, a little biphenyl is added to the mixture.) The cooling bath is removed, and the mixture, a dark-red suspension, is stirred for a further 15 min. The solution is decanted through glass wool to remove the excess of lithium. The yield of allyllithium is ca. 65%.

4-t-Butyl-1-phenylthiocyclohexyllithium (10)

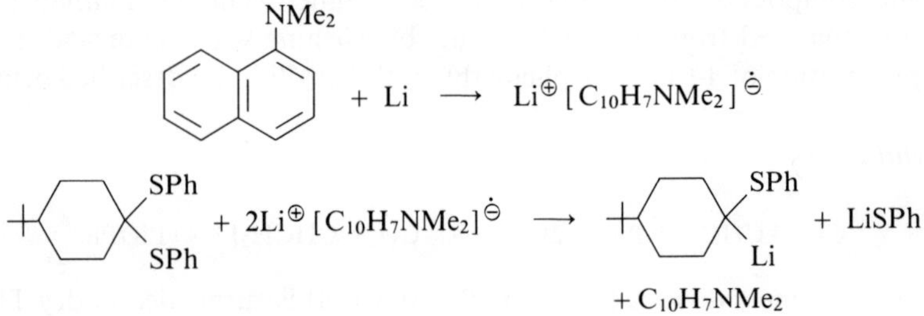

$$\text{C}_{10}\text{H}_7\text{NMe}_2 + \text{Li} \longrightarrow \text{Li}^\oplus [\text{C}_{10}\text{H}_7\text{NMe}_2]^{\ominus\cdot}$$

$$\text{Bu}^t\text{-C}_6\text{H}_9(\text{SPh})_2 + 2\text{Li}^\oplus [\text{C}_{10}\text{H}_7\text{NMe}_2]^{\ominus\cdot} \longrightarrow \text{Bu}^t\text{-C}_6\text{H}_9(\text{SPh})(\text{Li}) + \text{LiSPh} + \text{C}_{10}\text{H}_7\text{NMe}_2$$

3.5. FROM SULPHONYLHYDRAZONES

A dry two-necked flask is fitted with a septum and a glass-coated magnetic stirrer bar, and furnished with an atmosphere of argon. THF (20 ml) and clean lithium ribbon (80 mg, 11.6 mmol)* are added. The mixture is cooled in a bath at $-45°$ to $-55°$ and stirred as 1-dimethylaminonaphthalene (1.68 ml, 1.74 g, 10.2 mmol) is slowly added. Within 10 min a dark green colour should develop. The mixture is stirred rapidly for 3.5 h.

The mixture is cooled to $-78°$ and a solution of 4-t-butyl-1,1-bis(phenylthio)cyclohexane (1.44 g, 4.05 mmol) in THF (5 ml) is added. The mixture is stirred for 15 min.

In the experiment described (10) trimethylsilyl chloride was added. The work-up involved immediate quenching with water at —78°, and extraction of an ether solution of the products with dilute alkali and dilute acid. 4-t-Butyl-1-phenylthio-1-trimethylsilylcyclohexane was obtained in 83% yield.

References

1. J. J. Eisch, *Organomet. Synth.* **2**, 91 (1981).
2. J. J. Eisch, *Organomet. Synth.* **2**, 95 (1981).
2a. A. Maercker, *Angew. Chem. Int. Ed. Engl.* **26**, 972 (1987).
3. J. A. Katzenellenbogen and R. S. Lenox, *J. Org. Chem.* **38**, 326 (1973).
4. C. G. Screttas and M. Micha-Screttas, *J. Org. Chem.* **43**, 1064 (1978).
5. C. G. Screttas and M. Micha-Screttas, *J. Org. Chem.* **44**, 713 (1979).
6. T. Cohen and R. B. Weisenfeld, *J. Org. Chem.* **44**, 3601 (1979).
7. T. Cohen and B.-S. Guo, *Tetrahedron* **42**, 2803 (1986).
8. C. Rücker, *J. Organomet. Chem.* **310**, 135 (1986).
9. J. J. Eisch and A. M. Jacobs, *J. Org. Chem.* **28**, 2145 (1963).
10. T. Cohen, J. P. Sherbine, J. R. Matz, R. R. Hutchins, B. M. McHenry and P. R. Willey, *J. Am. Chem. Soc.* **106**, 3245 (1984).

3.5. PREPARATION FROM SULPHONYLHYDRAZONES

The reaction of ketone arenesulphonylhydrazones with an excess of an organolithium compound was originally exploited, notably by Shapiro, as a synthesis of alkenes (1); the reaction is now commonly known as the Shapiro reaction. When the mechanism of the reaction was investigated, it was demonstrated that alkenyllithium compounds were intermediates, and the reaction is now regarded as a valuable synthesis of 1-alkenyllithium compounds:

$$R-C(=O)-R' \xrightarrow{ArSO_2NHNH_2} R-C(=NNHSO_2Ar)-R' \xrightarrow{2R''Li} R-C(Li)=CH-R'$$

*The presence of a freshly exposed clean lithium surface is critical.

Tosylhydrazones may be used, but are susceptible to metallation of the ring or the methyl group, so 2,4,6-triisopropylbenzenesulphonylhydrazones are employed for maximum yields. In the *Organic Syntheses* paper on which the following description is based, the range of ketones that may be used, and the modifications that may be adopted, are briefly reviewed (*2*). A good example of the application of the method to an αβ-unsaturated ketone is the synthesis of 2-lithio-1,3-butadiene (*3*).

2-Lithiobornene (*2*)

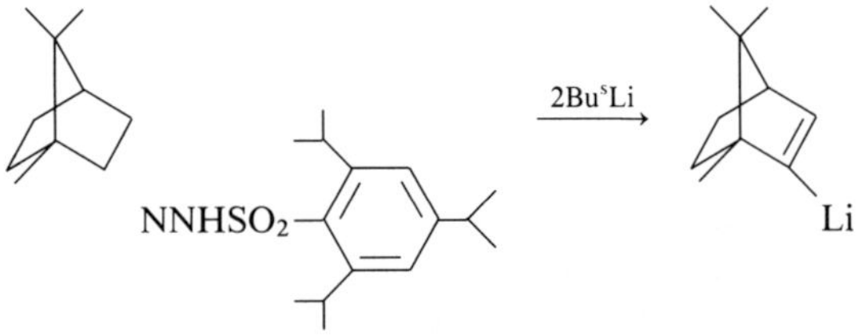

A 1 l three-necked flask is equipped with a septum, a stirrer and a 250 ml pressure-equalizing dropping funnel sealed with a septum, and furnished with an atmosphere of nitrogen. d-Camphor 2,4,6-triisopropylbenzene-sulphonylhydrazone (40.0 g, 92 mmol), hexane (200 ml) and TMEDA (200 ml) are added. The solution is cooled in a bath at *ca.* − 55° and stirred rapidly while s-butyllithium (*ca.* 1.3 M in hydrocarbon, 0.20 mol) is added from the funnel during *ca.* 20 min. The resulting orange solution is stirred and cooled for 2 h. The cooling bath is removed, and after 20 min replaced by an ice bath. Nitrogen evolution continues for *ca.* 10 min.

In the experiment described, further reaction with 1-bromobutane gave 2-butylbornene (50–53%) together with bornene (*ca.* 25%) (*2*).

References

1. R. H. Shapiro, *Org. React.* **23**, 405 (1976).
2. A. R. Chamberlin, E. L. Liotta and F. T. Bond, *Org. Synth.* **61**, 141 (1982); for another example see N. E. Schore and M. J. Knudsen, *J. Org. Chem.* **52**, 569 (1987).
3. P. A. Brown and P. R. Jenkins, *J. Chem. Soc. Perkin Trans. 1* 1129 (1986).

3.6. PREPARATION BY ADDITION TO CARBON–CARBON AND CARBON–SULPHUR MULTIPLE BONDS

The addition of organolithium compounds to carbon–carbon double and triple bonds is covered in Chapter 4, and thiophilic addition to thiocarbonyl groups is covered in Section 7.2.

—4—
Addition of Organolithium Compounds to Carbon–Carbon Multiple Bonds

It is paradoxical that while the major industrial application of organolithium compounds—initiation of diene polymerization—involves their addition to carbon–carbon multiple bonds, such reactions are of only limited usefulness for laboratory-scale organic syntheses (see General Refs A and D). Nevertheless, there are certain categories of alkenes and alkynes that do undergo ready addition: strained alkenes, alkenyl and alkynyl derivatives of second- and third-row elements, and alkenes bearing electron-donor groups suitably placed to provide intramolecular assistance. Examples of the last two categories are descibed below. *Conjugated* carbon–carbon double and triple bonds also often readily undergo addition of organolithium compounds, as in the initiation of polymerization of dienes and styrenes (see General Refs A and D(ii) and Ref. (*1*)). Conjugate addition to $\alpha\beta$-unsaturated carbonyl compounds etc. is described in connection with the appropriate functional groups.*

Addition of organolithium compounds to arenes has been observed, as has subsequent elimination of lithium hydride to give overall alkylation. However, such reactions are again of limited preparative value. On the other hand, addition to coordinated arenes has promise as an indirect method for achieving overall aromatic nucleophilic substitution with formal displacement of hydride (*2*). In the example given below (*3*), the nucleophile is a lithiated cyanohydrin derivative and the overall reaction is equivalent to nucleophilic acylation of the aromatic ring. Various extensions of this type of reaction are possible, such as the following (*4*):

---

*Conjugate addition to $\alpha\beta$-unsaturated sulphones, a type of reaction not noted elsewhere, has proved useful (*5*), but displacement of the sulphonyl group and/or metallation are alternative reactions (*6*).

Mention should also be made of the addition of organocopper compounds to alkynes, since, although most work has been carried out with organocopper reagents derived from Grignard reagents, those derived from organolithium compounds may also be employed (7).

2-(2,2-Dimethylpropyl)-2-lithio-1,3-dithiane by addition of t-butyllithium to 2-methylene-1,3-dithiane (8)

$$\begin{array}{c} \diagup S \\ \diagdown S \end{array}\!\!=\!\!\!\!\!\!\!\!\!\xrightarrow{Bu^tLi}\!\!\!\!\!\!\!\begin{array}{c} \diagup S \\ \diagdown S \end{array}\!\!\!\!\!\!\begin{array}{c} Bu^t \\ \diagdown Li \end{array}$$

2-Methylene-1,3-dithiane (3.96 g, 30 mmol) is dissolved in dry THF (60 ml) under an atmosphere of argon and the solution is cooled to $-78°$. t-Butyllithium (ca. 1.6 M in pentane, 33 mmol) is added dropwise with stirring. The resulting yellow solution is stirred at $-20°$ for 3–4 h.

Addition of iodomethane (4.3 g, 30.3 mmol) to a solution thus prepared gave, after conventional work-up, 2-(2,2-dimethylpropyl)-2-methyl-1,3-dithiane (89%), from which 2,2-dimethylpentan-4-one could be obtained by treatment with mercury(II) oxide and chloride in methanol (8).

Addition of n-butyllithium to allyl alcohol; 2-methylhexan-1-ol (9)

$$\diagup\!\!\!\diagup\!\!\!\diagdown OH \xrightarrow[TMEDA]{2Bu^nLi} Li\diagdown\!\!\!\diagup\!\!\!\diagup\!\!\!\overset{Bu^n}{|}\!\!\!\diagdown OLi \xrightarrow{H_2O} \diagup\!\!\!\overset{Bu^n}{|}\!\!\!\diagdown OH$$

A dry 500 ml three-necked flask is fitted with a gas inlet, a septum and a reflux condenser, equipped for stirring, and furnished with a nitrogen atmosphere. The flask is cooled in ice as allyl alcohol (prop-2-en-1-ol) (7.25 g, 0.125 mol), pentane[a] (70 ml) and TMEDA (1.16 g, 0.010 mol) are successively added via the septum. The reaction mixture is well stirred as n-butyllithium (ca. 1.5 M in pentane,[a] 0.270 mol) is added via the septum during 20 min. The ice bath is removed and the mixture is stirred for 1 h. The ice bath is restored, the gas inlet is replaced by a pressure-equalizing dropping funnel, and water (70 ml) is added (CAUTION: exothermic reaction).

The organic layer is separated, washed with 3 M HCl (2 × 10 ml) and water (2 × 10 ml), and dried (MgSO$_4$). Most of the solvent is removed by distillation through a fractionating column and the residue is distilled to yield 2-methylhexan-1-ol (64–74%),[b] b.p. 166–167°.

4. ADDITION TO C–C MULTIPLE BONDS

[a] Hexane could be used, but would make separation of the solvent from the product more difficult.
[b] GLC revealed the presence of a trace of heptan-3-ol impurity (8).

Reaction of lithiated O-(1-ethoxyethyl) 2-methylpropanal cyanohydrin with (η^6-2-methylanisole)tricarbonylchromium; 2-methyl-5-(2-methylpropanoyl)anisole

This procedure, contributed by Semmelhack (3), illustrates the use of nucleophilic addition to an arenetricarbonylchromium for the synthesis of aromatic compounds with unusual substitution patterns.

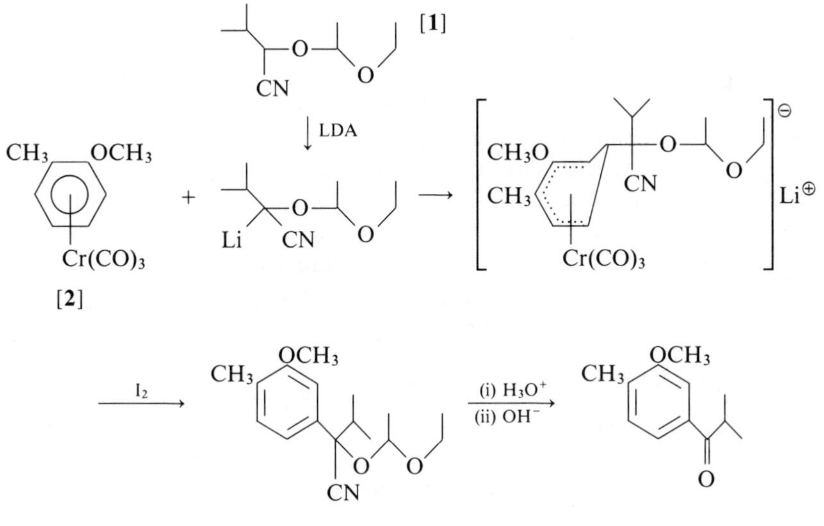

A 250 ml three-necked flask, equipped with a magnetic stirrer, a septum and a vacuum/argon inlet, is evacuated by means of an oil pump and refilled with argon three times. Diisopropylamine (dried by passage through a short column of basic alumina, activity grade I) (1.68 ml, 1.21 g, 12.0 mmol) and THF (distilled from benzophenone ketyl under argon) (40 ml) are added by syringe and the mixture is cooled to −30°. n-Butyllithium (ca. 1.5M in hexane, 12.0 mmol) is added dropwise during 2 min, and the resulting mixture is stirred at −30° for 45 min. The solution is cooled to −78° and a solution of the cyanohydrin acetal [1] (2.02 g, 12.0 mmol) in hexametapol (distilled from calcium hydride under argon) (10 ml) is added dropwise during 2 min; during the addition the solution turns red. After addition is complete, the solution is stirred for 1 h at −78°. A solution of the arene complex [2] (3.1 g, 12.0 mmol) in THF (5 ml) is added rapidly. The mixture is warmed to 0° slowly during 0.5 h and stirred at this temperature for 1 h.

It is then cooled to −78° and a solution of iodine (12 g) in THF (45 ml) is added. The resulting dark-brown solution is stirred for 4 h at 25° and then poured into ether (500 ml). The ether solution is washed with saturated aqueous sodium bisulphite (100 ml) to reduce the excess of iodine, and the red-brown precipitate that appears is removed by filtration. The ether layer is washed sequentially with saturated aqueous sodium bisulphite (100 ml), water (100 ml), saturated aqueous sodium bicarbonate (100 ml) and saturated aqueous sodium chloride (100 ml). It is dried ($MgSO_4$), filtered, and concentrated by rotary evaporation. The residual liquid is dissolved in methanol (50 ml) and 5% aqueous sulphuric acid is added in order to cleave the acetal protecting group. After having been stirred at 25° for 0.5 h, the solution is neutralized by addition of solid sodium bicarbonate and concentrated in a rotary evaporator to remove the methanol and most of the water. The residue, a solid mass, is triturated with ether (total of 500 ml) and the ether solution is shaken vigorously with 5% aqueous sodium hydroxide (100 ml) to regenerate the ketone from the cyanohydrin. The ether layer is washed sequentially with saturated aqueous sodium chloride (100 ml), 5% hydrochloric acid (100 ml), and saturated aqueous sodium bicarbonate (100 ml). It is dried ($MgSO_4$), filtered, and concentrated by rotary evaporation to leave a yellow liquid residue. Bulb-to-bulb distillation (oven temperature 110–130°, 0.01 Torr) produces as a colourless liquid 2-methyl-5-(2-methylpropanoyl)anisole (2.06 g, 91%).[a]

[a] The product showed no significant impurities by ^{1}H NMR analysis. Analytically pure material was obtained by simple distillation (b.p. 88–96°/0.02 Torr) of a sample from a similar experiment (3).

REFERENCES

1. M. Szwarc (ed.), *Ions and Ion Pairs in Organic Reactions*, Vols 1 and 2, Wiley-Interscience, New York, 1972 and 1974; see especially Vol. 1, Chap. 4.
2. M. F. Semmelhack, *New Applications of Organometallic Reagents in Organic Synthesis*, ed. D. Seyferth, Elsevier, Amsterdam, 1978, p. 361.
3. M. F. Semmelhack, Personal communication.
4. J. Blagg, S. G. Davies, C. L. Goodfellow and K. E. Sutton, *J. Chem. Soc. Chem. Commun.* 1283 (1986).
5. P. C. Conrad, P. L. Kwiatkowski and P. L. Fuchs, *J. Org. Chem.* **52**, 586 (1987) and refs therein.
6. See e.g. J. J. Eisch, M. Behrooz and J. E. Galle, *Tetrahedron Lett.* **25**, 4851 (1984); A. Padwa and M. W. Wannamaker, *Tetrahedron Lett.* **27**, 5817 (1986); M. Yamamoto, K. Suzuki, S. Tanaka and K. Yamada, *Bull. Chem. Soc. Jpn* **60**, 1923 (1987).
7. A. Alexakis and J. F. Normant, *Tetrahedron Lett.* **23**, 5151 (1982); M. Furber, R. J. K. Taylor and S. G. Burford, *Tetrahedron Lett.* **26**, 3285 (1985).
8. D. Seebach, R. Bürstinghaus, B.-T. Gröbel and M. Kolb, *Justus Liebigs Ann. Chem.* 830 (1977).
9. J. K. Crandall and A. J. Rojas, *Org. Synth.* **55**, 1 (1976).

—5—
Addition of Organolithium Compounds to Carbon–Nitrogen Multiple Bonds

5.1. ADDITION TO IMINES AND IMINIUM SALTS

The addition of organolithium compounds to simple imines is much less satisfactory as a general synthetic method than might be supposed, particularly when α-hydrogens are present (see General Ref. D(ii)):

$$R^1R^2C=NR^3 \xrightarrow{R^4Li} R^1R^2R^4C-N(Li)R^3 \xrightarrow{H^+} R^1R^2R^4C-NHR^3$$

A possible way of minimizing α-deprotonation is by converting the organolithium compound into a cuprate *in situ*, and activating the cuprate with boron trifluoride (*1*).

Reactions of organolithium compounds with $\alpha\beta$-unsaturated imines (1,2- and/or 1,4-addition (see General Ref. D(ii))), aryl aldimines (see General Ref. D(ii) and Ref. (*2*)) and N-alkenylimines (leading to lithioenamines) (*3*) are more useful. There have also been a few reports of useful reactions involving addition to the azomethine bond of compounds such as hydrazones (see General Ref. D(ii)) and carbodiimides (*4*):

$$R^1CH=NNHR^2 \xrightarrow[\text{(ii) } H^+]{\text{(i) } R^3Li} R^1R^3CHNHNHR^2$$

$$R^1N=C=NR^1 \xrightarrow[\text{(ii) } H^+]{\text{(i) } R^2Li} R^1N=\overset{R^2}{\underset{|}{C}}-NHR^1$$

Addition of organolithium compounds to iminium salts has been little explored (see General Ref. D(ii)), but may well be less susceptible to side-reactions than addition to imines. Such reactions merit further study (see e.g. Ref. (*5*)), so a simple example is described.

Addition of n-butyllithium to dimethyl(methylene)ammonium iodide (Eschenmoser's salt) (6)

$$CH_2=N^{\oplus}Me_2 \ I^{\ominus} \xrightarrow{Bu^nLi} BuCH_2NMe_2 \xrightarrow{MeI} BuCH_2N^{\oplus}Me_3 \ I^{\ominus}$$

Eschenmoser's salt (2.04 g, 11 mmol) is suspended in dry ether (50 ml) in a flask fitted with a gas inlet, a septum and a reflux condenser, and furnished with a stirrer and a nitrogen atmosphere. n-Butyllithium (*ca.* 2.4 M in hexane, 10 mmol) is added via syringe to the stirred suspension at a rate which maintains a gentle reflux. When the addition is complete the mixture is stirred vigorously at room temperature for 1 h. Water is added with care until the suspended salts just dissolve. The mixture is stirred vigorously for 30 min. The ether layer is separated, and the aqueous layer is extracted three times with ether. The combined ether layers are washed with saturated brine, dried (MgSO$_4$) and concentrated. An excess of methyl iodide is added to the remaining ether solution. After several hours the resulting suspension is filtered, and the solid, trimethyl(pentyl)ammonium iodide (84%), is washed with ether and dried.

References

1. M. Wada, Y. Sakurai and K.-Y. Akiba, *Nippon Kagaku Kaishi* 295 (1985).
2. I. A. Cliffe, R. Crossley and R. G. Shepherd, *Synthesis* 1183 (1985).
3. P. A. Wender and M. A. Eissenstat, *J. Am. Chem. Soc.* **100**, 292 (1978).
4. J. Pornet and L. Miginiac, *Bull. Soc. Chim. Fr.* 994 (1974).
5. Jahangir, D. B. MacLean, M. A. Brook and H. L. Holland, *J. Chem. Soc. Chem. Commun.* 1608 (1986); *Tetrahedron*, **43**, 5761 (1987).
6. J. L. Roberts, P. S. Borromeo and C. D. Poulter, *Tetrahedron Lett.* 1299 (1977).

5.2. ADDITION TO NITROGEN HETEROCYCLIC AROMATIC COMPOUNDS

The addition of an organolithium compound to the formal azomethine link of pyridine or another nitrogen aromatic heterocycle is a well-established general reaction (see General Refs A and D(ii)). It is usually used to achieve overall substitution on the ring, via lithium hydride elimination or oxidation of the dihydro-intermediate, but other reactions of the initial adduct are also useful. Some reactions involving an initial addition of phenyllithium to pyridine are described below. Some well-described examples of additions to other heterocycles are listed in Table 5.1.

TABLE 5.1

Addition of Organolithium Compounds to Nitrogen Heterocyclic Aromatic Compounds

Heterocyclic compound	Organolithium compound (solvent)	Product (yield %)	Ref.
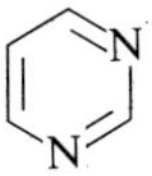 pyrimidine	 2-thienyllithium (Et$_2$O + hexane)	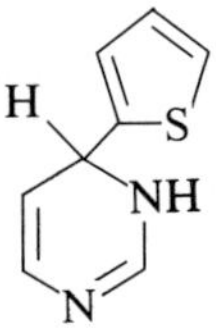 4-(2-thienyl)-3,4-dihydropyrimidine (65)[a]	(3)
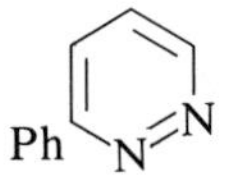 3-phenylpyridazine	PhLi (Et$_2$O + benzene)	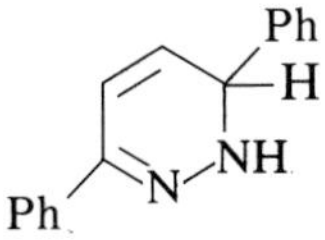 3,6-diphenyl-2,3-dihydropyridazine (—)[b]	(4)
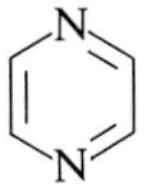 pyrazine	PhLi (THF)	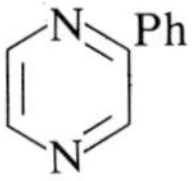 2-phenylpyrazine (60)[c]	(4)
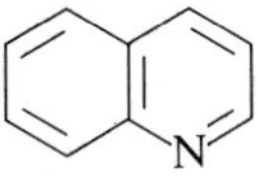 quinoline	Bu^{n}Li (benzene)	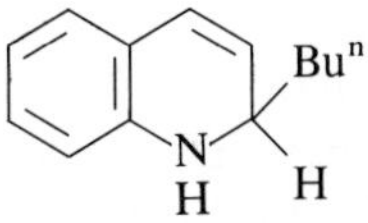 2-n-butyl-1,2-dihydroquinoline (90)	(5)
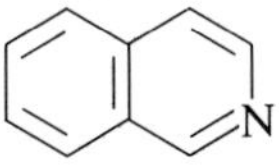 isoquinoline	Bu^{n}Li (benzene)	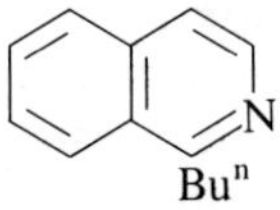 1-n-butylisoquinoline (70)	(5)
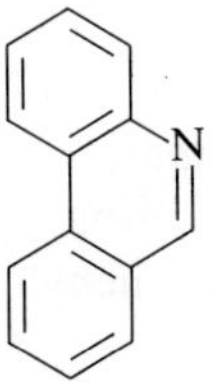 phenanthridine	Bu^{n}Li (Et$_2$O)	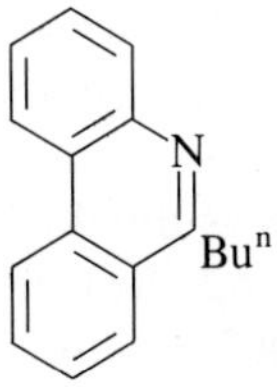 6-n-butylphenanthridine (90)[d]	(6)
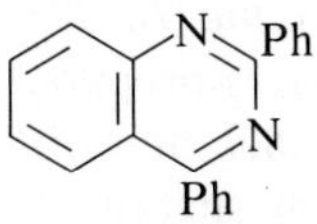 2,4-diphenylquinazoline	MeLi (Et$_2$O)	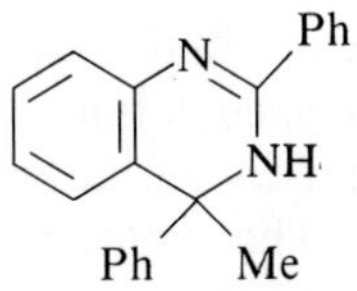 4-methyl-2,4-diphenyl-3,4-dihydroquinazoline (55)	(7)

[a] Or 3,4-dihydro-isomer; cf. Ref. (4). [b] Not stated; see also Ref. (8). [c] After oxidation by air. [d] After oxidation by nitrobenzene; still contained a little dihydro.

In general, attack occurs at the ring positions indicated in Table 5.1 (and a more complete list is given in General Ref. D(ii)). The effect of simple substituents on the position of attack is often not clear-cut (1), but strongly coordinating substituents can have a profound effect. For example, a 3-oxazinyl group directs attack by many organolithium compounds mainly to the 4-position of pyridine (2).

Reactions proceeding via addition of phenyllithium to pyridine

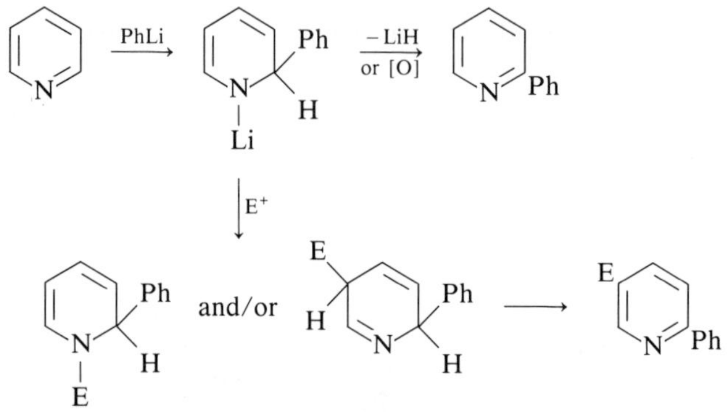

(i) 2-Phenylpyridine (9). A solution of phenyllithium in diethyl ether (150 ml) is prepared as described on p. 24 from bromobenzene (40 g, 0.25 mol) in a 1 l three-necked flask fitted with a dropping funnel, reflux condenser and stirrer, and flushed with nitrogen. The solution is stirred as a solution of *dry* pyridine* (40 g, 0.5 mol) in dry toluene (100 ml) is added slowly. Distillation apparatus is fitted over the reflux condenser. Water is run out of the reflux condenser, and solvent is distilled until the internal temperature reaches 110°. The water is reconnected to the reflux condenser and the reaction mixture is heated under reflux for *ca.* 8 h. It is then allowed to cool to *ca.* 40° and water (35 ml) is added cautiously. The layers are separated. The toluene layer is dried (KOH) and fractionally distilled, first at atmospheric pressure and then under reduced pressure. The yield of 2-phenylpyridine, b.p. 140°/12 mmHg is up to *ca.* 20 g (52%).

(ii) Isolation and reactions of 1-lithio-2-phenyl-1,2-dihydropyridine (10, 11). A solution of phenyllithium (0.01 mol) in diethyl ether (20 ml) is prepared (see p. 24) and cooled in an ice-water bath. Pyridine (0.7 g, 8.7 mmol) is added slowly. The resulting yellow precipitate is redissolved by shaking or stirring, and the mixture is allowed to stand at 0° until crystallization is

* Distilled from potassium hydroxide and/or calcium or barium oxide.

5.2. ADDITION TO AROMATIC N-HETEROCYCLES

TABLE 5.2
Reactions of 1-Lithio-2-phenyl-1,2-dihydropyridine

Reagent	Product	Yield (%)	Ref.
MeCOCl	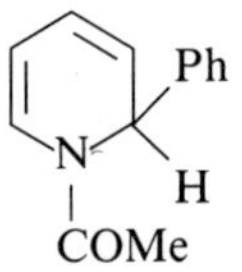 2-Ph, N-COMe dihydropyridine	68	(10)
ClCO$_2$Et	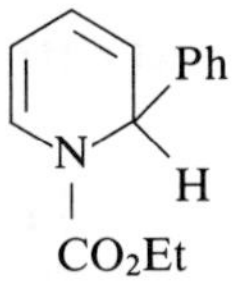 2-Ph, N-CO$_2$Et dihydropyridine	77	(10)
MeI	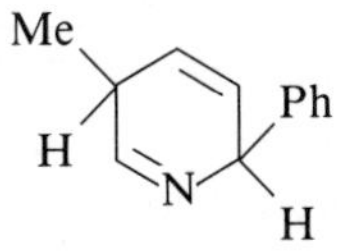 5-Me-2-Ph dihydropyridine (b)	(a)	(11)
EtBr	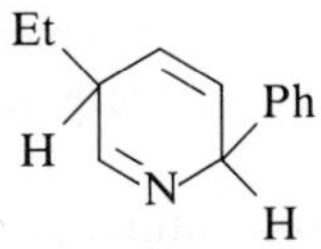 5-Et-2-Ph dihydropyridine (b)	—	(11)

^{a} Ref. (12) gives 34%. ^{b} On distillation, the 2,5-dialkyl-2,5-dihydropyridines gave the corresponding 2,5-dialkylpyridines. In some related cases disproportionation occurred, and further treatment with selenium was needed to obtain the fully aromatic product.

complete (ideally overnight). The supernatant liquid is removed by means of a syringe, and the crystals are washed by injecting dry pentane (10 ml), shaking the flask, allowing the solid to settle, and withdrawing the liquid.

Some reactions of the adduct thus prepared are summarized in Table 5.2. Experimental details (but not in all cases yields) are given in the references cited.

References

1. R. A. Abramovitch and J. G. Saha, *Adv. Heterocycl. Chem.* **6**, 229 (1966).
2. A. E. Hauck and C. S. Giam, *J. Chem. Soc. Perkin Trans. 1* 2227 (1984).
3. R. E. van der Stoel and H. C. van der Plas, *J. Chem. Soc. Perkin Trans. 1* 2393 (1979).
4. R. E. van der Stoel and H. C. van der Plas, *Recl. Trav. Chim. Pays-Bas* **97**, 116 (1978).
5. K. Ziegler and H. Zeiser, *Justus Liebigs Ann. Chem.* **485**, 174 (1931).

6. H. Gilman and J. Eisch, *J. Am. Chem. Soc.* **79**, 4423 (1957).
7. J. G. Smith and J. M. Sheepy, *J. Heterocycl. Chem.* **12**, 231 (1975); see also F. Johannsen and E. B. Pedersen, *Chem. Scripta* **27**, 277 (1987).
8. R. E. van der Stoel, H. C. van der Plas, H. Jongejan and L. Hoeve, *Recl. Trav. Chim. Pays-Bas* **99**, 234 (1980).
9. J. C. W. Evans and C. P. H. Allen, *Org. Synth. Coll. Vol.* **2**, 517 (1943).
10. C. S. Giam, E. E. Knaus and F. M. Pasutto, *J. Org. Chem.* **39**, 3565 (1974).
11. R. F. Francis, C. D. Crews and B. S. Scott, *J. Org. Chem.* **43**, 3227 (1978).
12. C. S. Giam and J. L. Stout, *Chem. Commun.* 478 (1970).

5.3. ADDITION TO NITRILES

The addition of organolithium compounds to nitriles is subject to various side reactions, notably α-deprotonation. The complexities are summarized in General Ref. D(ii). Reactions involving α-deprotonation may be useful in their own right (*1*), but when addition to the triple bond is desired they are sometimes difficult or impossible to avoid. Thus, while straightforward addition, giving the *N*-lithioketimine, predominates for aromatic nitriles, acetonitrile undergoes extensive deprotonation, and with phenylacetonitrile deprotonation is normally the main reaction. Barbier-type reactions between bromobenzene, lithium, and the nitrile, were ineffective for avoiding deprotonation (*2*).

Following formation of the *N*-lithioketimine, acid work-up is commonly used, giving a ketone, as in the synthesis of dicyclopentyl ketone described below. On milder hydrolysis the ketimine can be obtained, although, except where there is steric hindrance (as in the following example), special care may be needed to avoid further hydrolysis to the ketone. The *N*-lithioketimine may also be used for the synthesis of various *N*-substituted imines. The following is an example of an *intramolecular* reaction of the *N*-lithioimine (*3*):

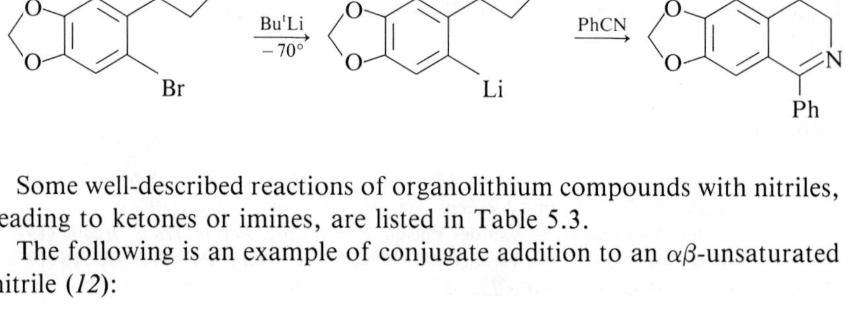

Some well-described reactions of organolithium compounds with nitriles, leading to ketones or imines, are listed in Table 5.3.

The following is an example of conjugate addition to an αβ-unsaturated nitrile (*12*):

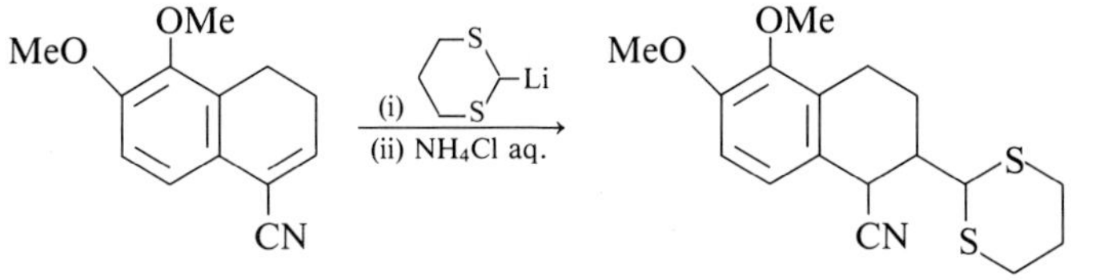

5.3. ADDITION TO NITRILES

[Reaction scheme: methoxy-substituted cyanonaphthalene + 1,3-dithianyl lithium, (i) then (ii) NH₄Cl aq. → tetrahydro adduct, 90%]

(ratio of *cis*:*trans* ca. 2.2:1)

TABLE 5.3
Reactions Involving Addition of Organolithium Compounds to Nitriles

Nitrile	Organo-lithium compound	Further reagent	Product (yield %)	Ref.
MeCN	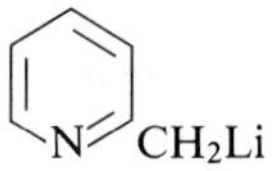 2-pyridyl-CH₂Li	H_2SO_4 aq.	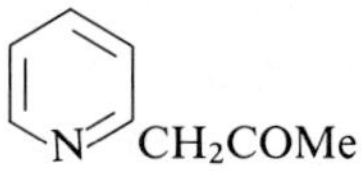 2-pyridyl-CH₂COMe (57)	(4)
EtCN	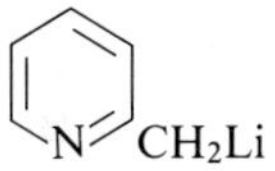 2-pyridyl-CH₂Li	H_2SO_4 aq.	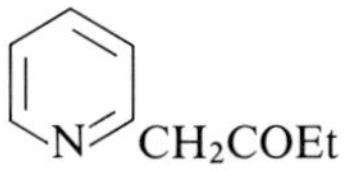 2-pyridyl-CH₂COEt (80)	(4)
$CH_3(CH_2)_4CN$	MeLi	H_2SO_4 aq.	$CH_3(CH_2)_4COMe$ (60)	(5)
$HC{\equiv}C(CH_2)_3CN$	2PhLi	H_2SO_4 aq.	$HC{\equiv}C(CH_2)_3COPh$ (60)	(6)
PhCN	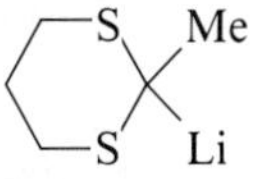 2-Me-1,3-dithian-2-yl-Li	HCl aq.	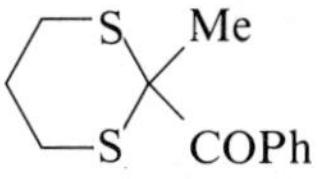 2-Me-2-(COPh)-1,3-dithiane (82)[a]	(7)
PhCN	 2-furyl-Li	HCl aq.	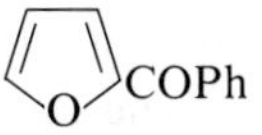 2-furyl-COPh (89)	(8)
PhCN	 2-thienyl-Li	HCl aq.	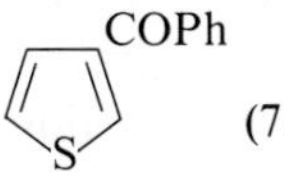 2-thienyl-COPh (75)	(9)
PhCN	C_6Cl_5Li	Me_2SO_4	$Ph(C_6Cl_5)C{=}NMe$ (55)	(10)
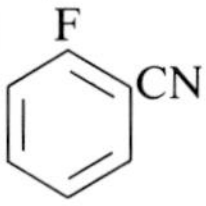 2-F-C₆H₄-CN	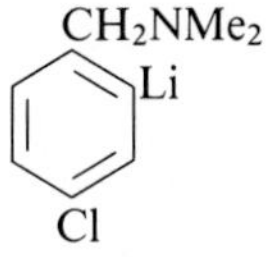 2-Cl-6-(CH₂NMe₂)-C₆H₃-Li	HCl aq.	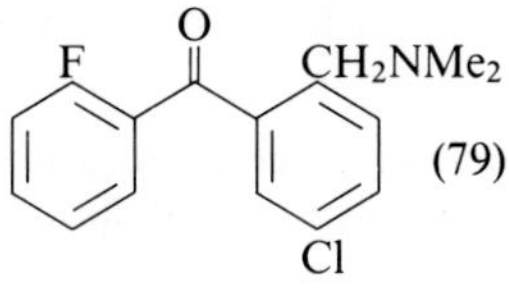 (2-F-C₆H₄)CO(2-Cl-6-(CH₂NMe₂)-C₆H₃) (79)	(11)

[a] Yield of crude intermediate imine, isolated before treatment with acid, was 95%.

Dicyclopentyl ketone (13)

$$\text{Cp-CN} + \text{Cp-Li} \longrightarrow [\text{Cp-C(=NLi)-Cp}] \xrightarrow{H_3O^+} \text{Cp-C(=O)-Cp}$$

A solution of cyclopentyllithium (1 mol) in cyclohexane is placed in a 2 l flask fitted with a stirrer and a pressure-equalizing dropping funnel and furnished with an atmosphere of nitrogen. Cyclohexane (500 ml) is added. The temperature is kept below 20° by external cooling as a solution of cyanocyclopentane (90 g, 0.95 mol) in cyclohexane (50 ml) is added dropwise, with stirring, during 30 min. The mixture turns bright yellow. Stirring is continued for 1 h. 10% hydrochloric acid (200 ml) is added, and the mixture is stirred rapidly at room temperature overnight. The organic layer is separated, washed with water (2 × 100 ml) and dried, and the solvent is evaporated. The residual yellow oil is distilled to give dicyclopentyl ketone (ca. 105 g, 66%), b.p. 70°/11 mmHg.

Di-t-butylketimine (14)

$$Bu^tLi + Bu^tCN \longrightarrow Bu^t_2C=NLi \xrightarrow{MeOH} Bu^t_2C=NH$$

t-Butyllithium (ca. 2 M in hexane,[a] 42 mmol) is introduced into a 100 ml flask equipped with a septum, a reflux condenser and a stirrer, and furnished with an inert atmosphere. The solution is frozen by cooling with liquid nitrogen, and a solution of t-butyl cyanide (3.48 g, 42 mmol) in pentane is added.[b] On warming to room temperature, a pale green-yellow solution is obtained.[c] The solution is re-cooled in liquid nitrogen. Methanol (5 ml) is added. The mixture is heated under reflux for 1 day,[d] cooled and filtered. The solvents are evaporated (bath temperature up to 140°) and the residue distilled, the imine being collected at 164–166°/755 mmHg.

Di-t-butylimides of beryllium (15), aluminium (16), boron (17), silicon (18), germanium (19) and tin (20) have been prepared by reactions of the N-lithioimine with the metal halides.

---

[a] Pentane or other hydrocarbon solvents may be used.
[b] Alternatively the t-butyllithium solution may be added to the frozen solution of the nitrile.
[c] The N-lithioketimine may be isolated by evaporation of the solvents.
[d] A much shorter time is probably sufficient.

References

1. S. Arseniyadis, K. S. Kyler and D. S. Watt, *Org. React.* **31**, 1 (1984).
2. G. G. Cameron and A. J. S. Milton, *J. Chem. Soc. Perkin Trans.* 2 378 (1976).
3. D. J. Jakiela, P. Helquist and L. D. Jones, *Org. Synth.* **62**, 74 (1984).
4. J. Büchi, F. Kracher and G. Schmidt, *Helv. Chim. Acta* **45**, 729 (1962).
5. G. Sumrell, *J. Org. Chem.* **19**, 817 (1954).
6. J. A. Gautier, M. Miocque and L. Mascrier-Demagny, *Bull. Soc. Chim. Fr.* 1551 (1967).
7. E. J. Corey and D. Seebach, *Angew. Chem. Int. Ed. Engl.* **4**, 1077 (1965).
8. V. Ramanathan and R. Levine, *J. Org. Chem.* **27**, 1216 (1962).
9. S. Gronowitz, *Arkiv Kemi* **12**, 533 (1958).
10. D. J. Berry and B. J. Wakefield, *J. Chem. Soc. (C)* 642 (1971).
11. See Section 3.2, Ref. (*3*).
12. F. Z. Basha, J. F. DeBernardis and S. Spanton, *J. Org. Chem.* **50**, 4160 (1985).
13. P. E. Eaton, C. Giordano, G. Schloemer and U. Vogel, *J. Org. Chem.* **41**, 2238 (1976).
14. C. Summerford, K. Wade and B. K. Wyatt, *J. Chem. Soc. (A)* 2016 (1970); W. Clegg, R. Snaith, H. M. M. Shearer, K. Wade and G. Whitehead, *J. Chem. Soc. Dalton Trans.* 1309 (1983).
15. B. Hall, J. B. Farmer, H. M. M. Shearer, J. D. Sowerby and K. Wade, *J. Chem. Soc. Dalton Trans.* 102 (1979).
16. R. Snaith, C. Summerford, K. Wade and B. K. Wyatt, *J. Chem. Soc. (A)* 2635 (1970).
17. M. R. Collier, M. F. Lappert, R. Snaith and K. Wade, *J. Chem. Soc. Dalton Trans.* 370 (1972).
18. J. B. Farmer, R. Snaith and K. Wade, *J. Chem. Soc. Dalton Trans.* 1501 (1972).
19. J. Keable, D. G. Othen and K. Wade, *J. Chem. Soc. Dalton Trans.* 1 (1976).
20. M. F. Lappert, J. McMeeking and D. E. Palmer, *J. Chem. Soc. Dalton Trans.* 150 (1972).

5.4. ADDITION TO ISONITRILES

Isonitriles possessing α-hydrogens tend to undergo deprotonation by organolithium compounds, rather than addition to the isocyano-group. In the absence of α-hydrogens, however, isonitriles show carbene-like reactivity, giving lithioimines; subsequent reaction with electrophiles gives imine derivatives, which may be desired in their own right or may be hydrolysed to the corresponding carbonyl compounds.

$$R-N=C: \xrightarrow{R'Li} R-N=C\begin{array}{c}R'\\ \diagdown\\ Li\end{array} \xrightarrow{E^+} R-N=C\begin{array}{c}R'\\ \diagdown\\ E\end{array} \xrightarrow{H_3O^+} O=C\begin{array}{c}R'\\ \diagdown\\ E\end{array}$$

In the latter case, the overall reaction is equivalent to nucleophilic acylation (*1*). The most convenient isonitrile for such reactions is 1,1,3,3-tetramethylbutylisonitrile.

4-Lithio-3,6,6,8,8-pentamethyl-5-azanon-4-ene (2)

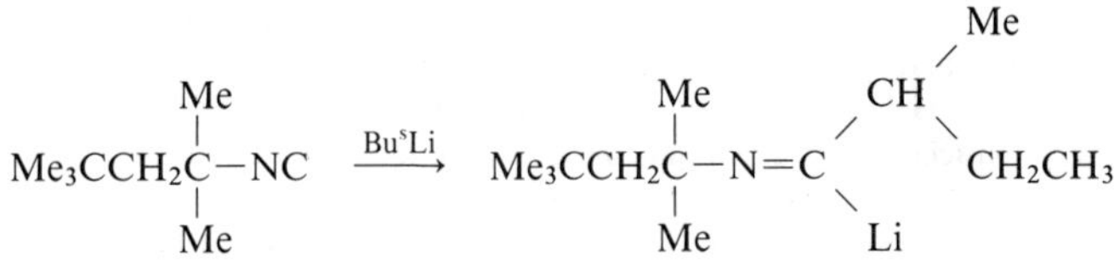

$$\text{Me}_3\text{CCH}_2\underset{\underset{\text{Me}}{|}}{\overset{\overset{\text{Me}}{|}}{\text{C}}}-\text{NC} \xrightarrow{\text{Bu}^s\text{Li}} \text{Me}_3\text{CCH}_2\underset{\underset{\text{Me}}{|}}{\overset{\overset{\text{Me}}{|}}{\text{C}}}-\text{N}=\text{C} \underset{\text{Li}}{\overset{\text{CH}\overset{\text{Me}}{\diagup}}{\diagdown \text{CH}_2\text{CH}_3}}$$

A dry 1 l three-necked flask is fitted with a mechanical stirrer, a 500 ml pressure-equalizing dropping funnel and a thermometer, and furnished with an atmosphere of nitrogen. A solution of 1,1,3,3-tetramethylbutylisonitrile (27.8 g, 35.1 ml, 0.2 mol)[a] in dry diethyl ether (300 ml) is added, and the solution is cooled to 0°. s-Butyllithium (in hexane, 0.2 mol) is transferred to the dropping funnel by syringe, and then added to the stirred solution at such a rate that the temperature does not rise above 5°; during the addition the mixture becomes gelatinous, and it is necessary to increase the stirring rate to ensure thorough mixing. After the addition is complete the mixture is stirred for a further 15 min at ca. 0°.

The addition of deuterium oxide to the reaction mixture to give N-(1-d-2-methylbutylidene)-1,1,3,3-tetramethylbutylamine (85–88%), and subsequent hydrolysis with aqueous oxalic acid to give 1-d-2-methylbutanal (87–88%), are described on p. 121.

[a] The synthesis of the isonitrile is also described (2).

References

1. G. E. Niznik, W. H. Morrison and H. M. Walborsky, *J. Org. Chem.* **39**, 600 (1974); N. Hirowatari and H. M. Walborsky, *J. Org. Chem.* **39**, 604 (1974); G. E. Niznik and H. M. Walborsky, *J. Org. Chem.* **39**, 608 (1974); M. Periasamy and H. M. Walborsky, *J. Org. Chem.* **39**, 611 (1974); H. M. Walborsky and P. Ronman, *J. Org. Chem.* **43**, 731 (1978).
2. G. E. Niznik, W. H. Morrison and H. M. Walborsky, *Org. Synth.* **51**, 31 (1971).

—6—
Addition of Organolithium Compounds to Carbonyl Groups

6.1. ADDITION TO ALDEHYDES AND KETONES

6.1.1. Addition to Saturated and Aryl Aldehydes and Ketones

In many cases the experimental procedures for reactions involving addition of organolithium compounds to aldehydes and ketones are straightforward, as in the following example and the ones listed in Table 6.1. Nevertheless, such reactions are susceptible to the same types of side-reactions as the analogous Grignard reactions: α-deprotonation (enolization) and reduction (both via β-hydrogen transfer and via ketyl formation).* Reduction is usually less troublesome in the case of organolithium compounds (though it may be significant with t-alkyl reagents). On the other hand, the high basicity of many organolithium compounds can make α-deprotonation of the carbonyl compound a real problem. In such cases, the use of a solvent of low polarity and/or the presence (or addition) of lithium salts is recommended (see General Ref. D(ii), p. 25). Recently, the conversion of an organolithium compound into an organotitanium reagent *in situ* has been used to avoid deprotonation, since the organotitanium reagents are only weakly basic, while retaining sufficient nucleophilicity (*11*). For example, β-tetralone undergoes extensive deprotonation by methyllithium, whereas a methyllithium–titanium tetrachloride reagent gives an almost quantitative yield of the addition product [1] (*11*):

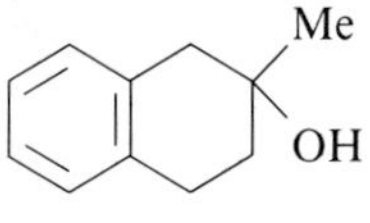

[1]

---

* In the course of a study of the mechanism of the addition reaction it was observed that a reaction of s-butyllithium with phenyl s-butyl ketone gave addition product (86%), reduction product (12%) and recovered ketone (presumably from enolization) (1%) (*10*).

TABLE 6.1

Addition of Organolithium Compounds to Saturated and Aryl Aldehydes and Ketones

Carbonyl compound	Organolithium compound	Product (yield %)	Ref.
$(CH_2O)_n$	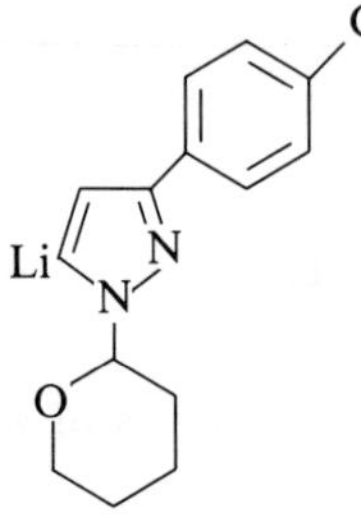 [Li-pyrazole with 4-Cl-phenyl and tetrahydropyranyl N-substituent]	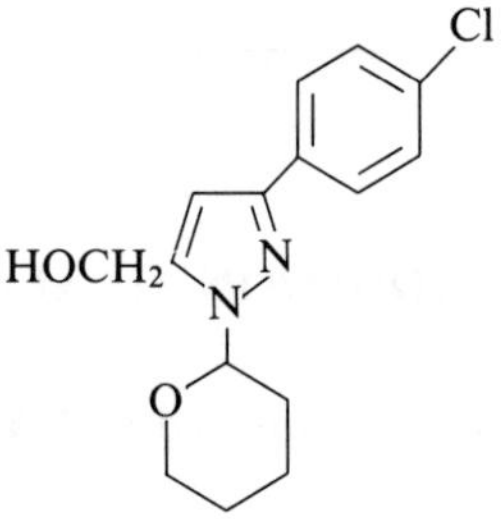 [HOCH$_2$-pyrazole with 4-Cl-phenyl and tetrahydropyranyl N-substituent] (72.5)	(1)
MeCHO	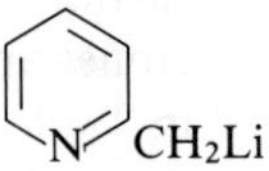 [2-pyridyl-CH$_2$Li]	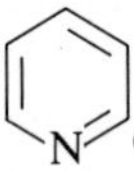 [2-pyridyl-CH$_2$CH(OH)Me] (44–50)	(2)
Me$_2$CHCH$_2$CHO	 [2-furyl-Li]	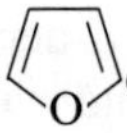 [2-furyl-CH(OH)CH$_2$CHMe$_2$] (93)	(3)
n-C$_5$H$_{11}$CHO	 [2-thienyl-Li]	 [2-thienyl-CH(OH)C$_5$H$_{11}$] (63)	(4)
PhCHO	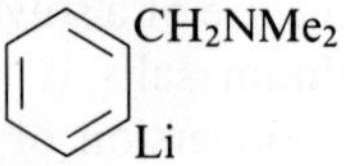 [o-(CH$_2$NMe$_2$)C$_6$H$_4$Li]	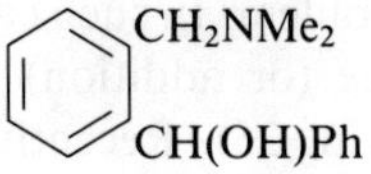 [o-(CH$_2$NMe$_2$)C$_6$H$_4$CH(OH)Ph] (70–73)	(5)
MeCOEt	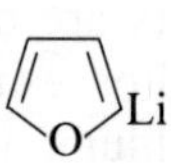 [2-furyl-Li]	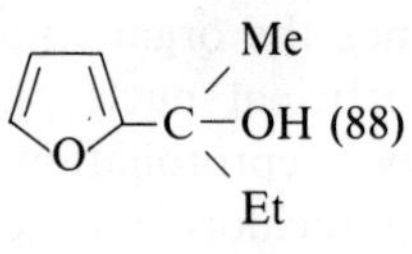 [2-furyl-C(Me)(Et)OH] (88)	(3)
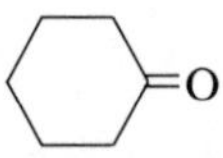 cyclohexanone	LiCH$_2$CO$_2$Et	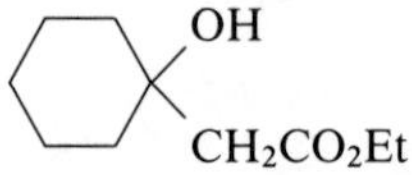 1-(CH$_2$CO$_2$Et)cyclohexan-1-ol (79–90)	(6)
Ph$_2$CO	MeLi	Ph$_2$CHMe^a (91–93)	(7)
Ph$_2$CO	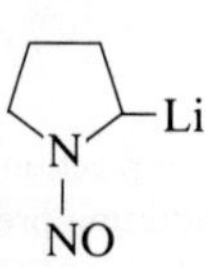 [1-(NO)pyrrolidin-2-yl-Li]	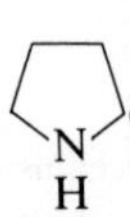 [1-H-pyrrolidin-2-yl-C(OH)Ph$_2$]^b (68–70)	(8)

6.1. ADDITION TO ALDEHYDES AND KETONES

TABLE 6.1—continued

Ph$_2$CO	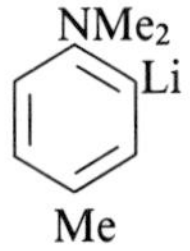 ⟨NMe$_2$, Li⟩ on benzene ring with Me	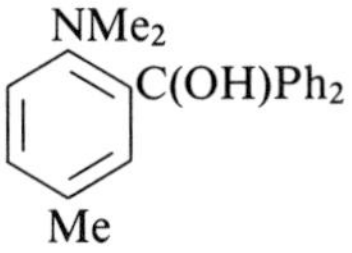 ⟨NMe$_2$, C(OH)Ph$_2$⟩ on benzene ring with Me	(49–57) (9)

^{a} After reduction *in situ*. ^{b} After denitrosation *in situ*.

The addition of cerium halides similarly suppresses enolization (*12*). Furthermore, the organotitanium reagents are selective in their reactivity towards different types of carbonyl group; they can differentiate aldehydes from ketones, as in the following example (*13*):*

$$PhCO(CH_2)_4CHO \xrightarrow{MeTi(OPr^i)_3} PhCO(CH_2)_4CH(OH)Me$$

Organohafnium compounds and organomanganese compounds are reported to show similar selectivity (*14*).

Just as the Barbier synthesis (*15*) has been revived as an alternative to the stepwise preparation of a Grignard reagent followed by reaction with a carbonyl compound, it has been found that a one-step reaction of lithium, an alkyl halide and a carbonyl compound can give good results, and indeed can reduce the amount of α-deprotonation. Many examples, with good experimental details, have been reported by Pearce, Richards and Scilly (*16*). The following preparation employed similar conditions.

A full discussion of the stereochemistry of the addition of organolithium compounds to carbonyl groups is beyond the scope of this book; the subject has been well reviewed (*17*). Nevertheless, reference should be made to asymmetric synthesis via additions of achiral organolithium compounds to achiral carbonyl compounds in the presence of chiral chelating ligands. The following successful example of this type of reaction has been described in *Organic Syntheses* (*18*):

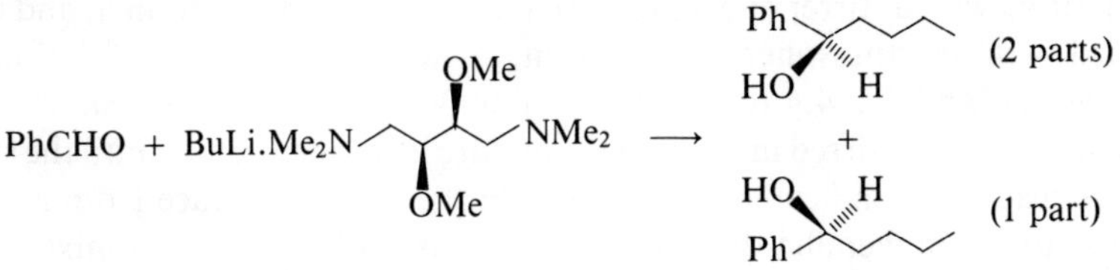

PhCHO + BuLi.Me$_2$N−CH(OMe)−CH(OMe)−NMe$_2$ → Ph−C(OH)(H)−Bu (2 parts) + Ph−C(OH)(H)−Bu (1 part)

*The reagent is prepared from methyllithium and chlorotriisopropoxytitanium and either isolated or used *in situ* (see p. 162).

6. ADDITION TO CARBONYL GROUPS

Addition of allyllithium to 4-methylpentan-2-one (19)

$$CH_2=CHCH_2Li + CH_3COCH_2CH(CH_3)_3 \longrightarrow CH_2=CHCH_2\underset{\underset{CH_3}{|}}{\overset{\overset{OLi}{|}}{C}}-CH_2CH(CH_3)_2$$

$$\xrightarrow{H_3O^+} CH_2=CHCH_2\underset{\underset{CH_3}{|}}{\overset{\overset{OH}{|}}{C}}-CH_2CH(CH_3)_2$$

A solution of allyllithium (0.12 mol) in diethyl ether (300 ml) is prepared in a 1 l three-necked flask fitted with a reflux condenser, a stirrer, a 250 ml pressure-equalizing dropping funnel, and flushed with nitrogen.[a] A solution of 4-methylpentan-2-one (12.0 g, 0.12 mol) in diethyl ether (25 ml) is placed in the dropping funnel and added at such a rate that the mixture boils gently. The mixture is heated under reflux for 1 h, then cooled to room temperature. Water (100 ml) is added.[b] The aqueous layer is extracted with ether (3 × 30 ml). The combined organic layers are dried (MgSO$_4$) and the ether is distilled at atmospheric pressure. The residue is filtered through sintered glass and fractionally distilled, the fraction b.p. 70–71°/20 mmHg being collected as 4,6-dimethylhept-1-en-4-ol (*ca.* 12.5 g, 70–75%).

[a] In the *Organic Syntheses* procedure the allyllithium is prepared from allyltriphenyltin and phenyllithium.
[b] In the *Organic Syntheses* procedure tetraphenyltin is removed at this point by filtration.

3-t-Butyl-3-hydroxy-2,2-dimethylheneicosane by a Barbier-type reaction (20)

$$(Me_3C)_2C=O + n\text{-}C_{18}H_{37}Cl \xrightarrow[\text{(ii) } H_3O^+]{\text{(i) Li, THF}} (Me_3C)_2C(OH)-n\text{-}C_{18}H_{37}$$

Lithium slices (3.46 g, 0.5 mol) are suspended in dry THF (120 ml) in a 500 ml flask fitted with a stirrer and a pressure-equalizing dropping funnel, and furnished with an atmosphere of nitrogen. A mixture (*ca.* 2 ml) of 1-chlorooctadecane and 2,2,4,4-tetramethylpentan-3-one (molar ratio *ca.* 2:1) is added to the well-stirred mixture. The mixture is warmed gently until the start of the reaction is indicated by a change in the lithium surface from a matt appearance to a metallic sheen. The mixture is cooled in ice as a mixture of 1-chlorooctadecane (69.4 g, 0.24 mol) and tetramethylpentanone (28.4 g, 0.20 mol) is added dropwise during 90 min. Stirring is continued for a further 1 h at 0°. The mixture is filtered, and the solvent is evaporated under reduced pressure. The residue is dissolved in light petroleum (b.p. 50–70°, 100 ml) and the solution is washed with 2 M hydrochloric acid and water and dried (potassium carbonate). The solvent is evaporated and the residue distilled in a molecular-distillation apparatus at 123–136°/10^{-5} Torr, giving 3-t-butyl-3-hydroxy-2,2-dimethylheneicosane (58.5 g, 74%), m.p. 28–30°.[a]

[a] After recrystallization, m.p. 34–35°.

References

1. See Section 3.2, Ref. (*3*).
2. L. A. Walter, *Org. Synth. Coll. Vol.* **3**, 757 (1955).
3. V. Ramanathan and R. Levine, *J. Org. Chem.* **27**, 1216 (1962).
4. G. Van Zyl, R. J. Langenberg, H. H. Tan and R. N. Schut, *J. Am. Chem. Soc.* **78**, 1955 (1956).
5. F. N. Jones, R. L. Vaulx and C. R. Hauser, *J. Org. Chem.* **28**, 3461 (1963).
6. M. W. Rathke, *Org. Synth.* **53**, 66 (1973).
7. S. D. Lipsky and S. S. Hall, *Org. Synth.* **55**, 7 (1975).
8. See Section 3.2, Ref. (*45*).
9. J. V. Hay and T. M. Harris, *Org. Synth.* **53**, 56 (1973).
10. M. A. Al-Aseer and S. G. Smith, *J. Org. Chem.* **49**, 2608 (1984).
11. M. T. Reetz, *Organotitanium Reagents in Organic Synthesis*, Springer-Verlag, Berlin, 1986.
12. T. Inamoto, T. Kusumoto, Y. Tawarayama, Y. Sugiura, T. Mita, Y. Hatanaka and M. Yokoyama, *J. Org. Chem.* **49**, 3904 (1984); L. A. Paquette and K. S. Learn, *J. Am. Chem. Soc.* **108**, 7873 (1986).
13. M. T. Reetz, J. Westermann, R. Steinbach, B. Wenderoth, R. Peter, R. Ostarek and S. Maus, *Chem. Ber.* **118**, 1421 (1985); M. T. Reetz and S. Maus, *Tetrahedron* **43**, 101 (1987); but see T. Kauffmann, T. Abel, M. Sahveer and D. Wingbermühle, *Tetrahedron* **43**, 2021 (1987).
14. T. Kauffmann, C. Pahde and D. Wingbermühle, *Tetrahedron Lett.* **26**, 4059 (1985); G. Cahiez and J. Normant, *Tetrahedron Lett.* 3383 (1977); G. Cahiez and B. Figadere, *Tetrahedron Lett.* **27**, 4445 (1986).
15. C. Blomberg and F. A. Hartog, *Synthesis* 18 (1977).
16. P. J. Pearce, D. H. Richards and N. F. Scilly, *J. Chem. Soc. Perkin Trans. 1* 1655 (1972).
17. J. D. Morrison (ed.), *Asymmetric Synthesis*, Vol. 2, Academic Press, New York, 1983.
18. D. Seebach and A. Hidber, *Org. Synth.* **61**, 42 (1983); see also M. D. Eleveld and H. Hogeveen, *Tetrahedron Lett.* **25**, 5187 (1984).
19. D. Seyferth and M. A. Weiner, *Org. Synth. Coll. Vol.* **5**, 452 (1973).
20. H. Quast and M. Heuschmann, *Synthesis* 117 (1976).

6.1.2. Addition to αβ-Unsaturated Aldehydes and Ketones

In general, organolithium compounds add to the carbonyl group of αβ-unsaturated carbonyl compounds, rather than giving 1,4 (Michael) addition (see General Refs A and D(ii)). The following example is fully described in

Organic Syntheses (1):

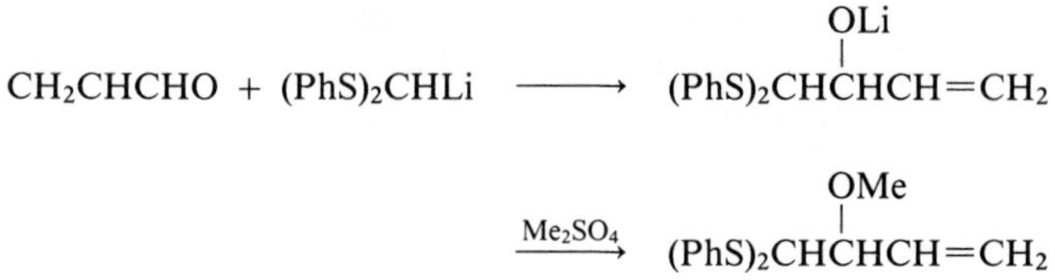

$$CH_2CHCHO + (PhS)_2CHLi \longrightarrow (PhS)_2CH\overset{\overset{\displaystyle OLi}{|}}{C}HCH=CH_2$$

$$\xrightarrow{Me_2SO_4} (PhS)_2CH\overset{\overset{\displaystyle OMe}{|}}{C}HCH=CH_2$$

The factors favouring 1,4-addition of organolithium compounds to $\alpha\beta$-unsaturated carbonyl compounds have been briefly reviewed (General Ref. D(ii), p. 28). They are: (i) solvents favouring electron transfer (notably hexametapol); (ii) delocalized organolithium compounds; and (iii) reaction times and/or temperatures allowing thermodynamic control.

Conjugate addition of α-lithio(4-methoxyphenyl)acetonitrile to cyclohex-2-enone (2)

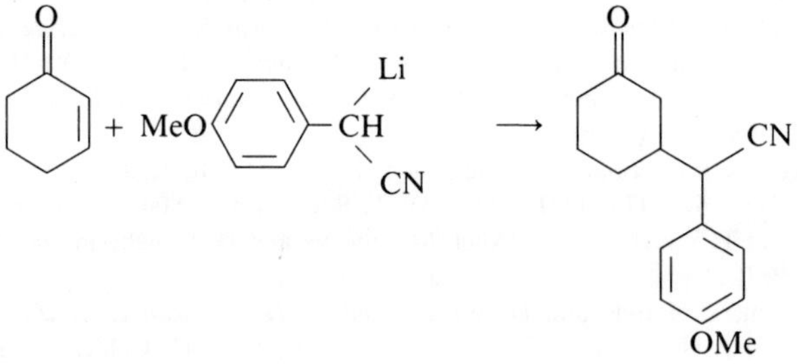

A dry four-necked flask is fitted with a mechanical stirrer, a gas inlet, a septum and a thermometer, and flushed with nitrogen or argon. THF (40 ml), hexametapol (10 ml), and 4-methoxyphenylacetonitrile (1.47 g, 10 mmol) are added by syringe and cooled to $-70°$. The solution is stirred at $-70°$ as n-butyllithium (in hexane, 10 mmol) is added by syringe. The solution is kept at $-70°$ as a solution of cyclohex-2-enone (0.96 g, 10 mmol) in THF (5 ml) is added, again by syringe. After a few min stirring, M HCl (20 ml) is added rapidly and the mixture is allowed to warm to room temperature. Ether is added, and the organic layer is separated, washed with saturated aqueous sodium chloride until the washings are neutral, and dried. Evaporation of the solvent gives, in virtually quantitative yield, 2-(4-methoxyphenyl)-2-(3-oxocyclohexyl)acetonitrile, which is purified by preparative TLC.

6.1. ADDITION TO ALDEHYDES AND KETONES

Conjugate additions of organocopper reagents to enones are well reviewed (3). Nevertheless, the organocopper reagents are usually prepared *in situ* from organolithium reagents, and such reactions are so useful that two examples are described below. One is a straightforward addition using a "higher-order" cyanocuprate (4). The other involves a further reaction of the enolate intermediate.

Addition of a cyanocuprate reagent; 4-methyl-4-phenylpentan-2-one (5)

$$Me_2C=CHCOMe \xrightarrow[\text{(ii) } H_2O]{\text{(i) } Ph_2Cu(CN)Li_2} Me_2\overset{\overset{\displaystyle Ph}{|}}{C}-CH_2COMe$$

Dry copper(I) cyanide (66 mg, 0.74 mmol) is placed in a dry two-necked flask fitted with a septum and a magnetic stirrer, and furnished with an atmosphere of argon. Ether (0.95 ml) is added, and the slurry is cooled to $-78°$ as phenyllithium (*ca.* 2.2. M in ether, 1.44 mmol) is added by syringe. Warming the mixture to $0°$ with gentle stirring produces a yellowish but not quite homogeneous solution. The mixture is re-cooled to $-78°$, and neat mesityl oxide (2-methylpent-2-en-3-one) (57 µl, 0.5 mmol) is added to the stirred mixture by syringe. Stirring is continued at $-78°$, though after 45 min the solution becomes viscous and further stirring is difficult. After 1 h, a 10% ammonium hydroxide/saturated aqueous ammonium chloride solution is added. The mixture is stirred for 10 min, and the crude product is isolated conventionally via ether extraction. Column chromatography (silica, 3:1 pentane–Et_2O) gives 4-methyl-4-phenylpentan-2-one (72.6 mg, 83%).

Addition of lithium dibutylcuprate to cyclohexenone, followed by reaction with iodomethane; 3-butyl-2-methylcyclohexanone (6)

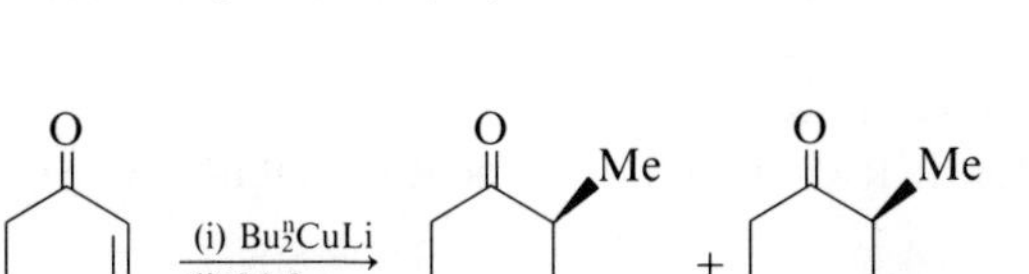

Copper(I) iodide (295 mg, 1.55 mmol) is suspended in THF (4 ml). The suspension is maintained at $-50°$ and stirred as n-butyllithium (*ca.* 2.5 M in hexane, 3.0 mmol) is added. The dark mixture is cooled to $-78°$ and stirred as a solution of cyclohex-2-enone (96 mg, 1.0 mmol) in THF (1 ml) is added. Stirring is continued for 30 min. A solution of iodomethane

(0.16 ml, 2.0 mmol) in hexametapol (2 ml) is added. The reaction mixture is allowed to warm to $-30°$ and maintained between $-30°$ and $-40°$ for 2 h. Methanol (1 ml) is added. The mixture is allowed to warm to room temperature, poured into saturated aqueous ammonium chloride, diluted with ether, and stirred for 1 h. The layers are separated, and the aqueous phase is extracted once with ether. The organic layers are combined, washed with 2% aqueous sodium thiosulphate, and dried over magnesium sulphate. Analysis by GLC shows the crude product to contain 3-butylcyclohexanone (1% yield), *trans*-3-butyl-2-methylcyclohexanone (74% yield), and *cis*-3-butyl-2-methylcyclohexanone (10% yield). The last two compounds may be isolated by preparative GLC.

References

1. See Section 3.2., Ref. (*47*).
2. M. C. Roux, L. Wartski and J. Seyden-Penne, *Tetrahedron* **37**, 1927 (1981).
3. G. H. Posner, *Org. React.* **19**, 1 (1972); G. H. Posner, *An Introduction to Organic Synthesis Using Organocopper Reagents*, Wiley, New York, 1980.
4. B. H. Lipshutz, R. S. Wilhelm and J. A. Kozlowski, *Tetrahedron* **40**, 5005 (1984); B. H. Lipshutz, *Synthesis* 325 (1987).
5. B. H. Lipshutz, R. S. Wilhelm and J. A. Kozlowski, *J. Org. Chem.* **49**, 3938 (1984).
6. G. H. Posner, J. J. Sterling, C. E. Whitten, C. M. Lentz and D. J. Brunelle, *J. Am. Chem. Soc.* **97**, 107 (1975).

6.1.3. Peterson Olefination and Related Reactions

The Peterson olefination reaction involves the addition of an α-silylorganolithium (or organomagnesium) compound to a carbonyl compound. The adduct can then eliminate lithium trialkylsilyloxide to give an alkene. Alternatively, hydrolysis may give an isolable β-hydroxysilane, which can be converted into an alkene by treatment with acid or base (*1, 2*).

$$R^1R^2\overset{Li}{C}-SiR_3^3 + R^4R^5C=O \longrightarrow \underset{R^1R^2C-SiR_3^3}{R^4R^5C-OLi} \xrightarrow{H_3O^+} \underset{R^1R^2C-SiR_3^3}{R^4R^5C-OH}$$

$$\downarrow {-R_3^3SiOLi} \qquad \diagup {H_3O^+ \text{ or } B^-}$$

$$R^4R^5C=CR^1R^2$$

Peterson olefination is in many ways complementary to the Wittig reaction (see Section 16.3), and, like the Wittig reaction, it is particularly valuable for the stereocontrolled synthesis of alkenes. The scope and stereochemistry of

6.1. ADDITION TO ALDEHYDES AND KETONES

the reaction have been reviewed (3). Various alternatives to the trialkylsilyl group have been reported (see General Ref. D(ii), p. 28), and the use of organolithium compounds in the Horner-Wadsworth-Emmons synthesis may be regarded as analogous (4):

$$R^1_2P(O)CH_2R^2 \xrightarrow{R^3Li} R^1_2P(O)\overset{Li}{\underset{|}{C}}HR^2 \xrightarrow[-R^1_2P(O)OLi]{R^4R^5C=O} R^2CH=CR^4R^5$$

Diphenyl(2,2-diphenylethenyl)phosphine sulphide

$$Ph_2P(S)CH_2SiMe_3 \xrightarrow{BuLi} Ph_2P(S)\overset{Li}{\underset{SiMe_3}{C}}H \xrightarrow{Ph_2CO} Ph_2P(S)CH=CPh_2$$

This procedure, based on Peterson's original paper (1), is for a reaction where the elimination occurs spontaneously, giving the olefin without isolation of the β-hydroxysilane.

A solution of diphenyl(trimethylsilylmethyl)phosphine sulphide (6.08 g, 20 mmol) in THF (30 ml) is prepared under an atmosphere of argon and cooled to 0°. n-Butyllithium (ca. 1.5 M in hexane, 21 mmol) is added dropwise with stirring. The mixture is stirred at 0° for 1.5 h and then added to a solution of benzophenone (3.64 g, 20 mmol) in THF (10 ml). After 0.5 h the reaction mixture is poured into aqueous ammonium chloride. The organic layer is separated and concentrated, and the resulting solid (ca. 7.5 g) is purified by chromatography on neutral alumina,* giving diphenyl(2,2-diphenylethenyl)phosphine sulphide (4.2 g pure, 1.3 g less pure, total 80%).

References

1. D. J. Peterson, *J. Org. Chem.* **33**, 780 (1968).
2. P. F. Hudrlik and D. Peterson, *J. Am. Chem. Soc.* **97**, 1464 (1975).
3. P. Magnus, T. Sarkar and S. Djuric, *Comprehensive Organometallic Chemistry*, ed. G. Wilkinson, Vol. 7, p. 520, Pergamon, Oxford, 1982; see also H. J. Reich, R. C. Holtan and S. L. Barkowsky, *J. Org. Chem.* **52**, 312 (1987) and C. R. Johnson and B. D. Tait, *Ibid.* p. 281.
4. W. J. Wadsworth, *Org React.* **25**, 73 (1977); see also A. D. Buss, N. Greaves, R. Mason and S. Warren, *J. Chem. Soc., Perkin Trans. 1*, 2569 (1987).

---

* Benzene was used as eluant (1). A less hazardous alternative would be preferable.

6.2. REACTIONS WITH ACYL HALIDES, ANHYDRIDES, ESTERS AND LACTONES

The reactions of organolithium compounds with acyl derivatives R^1COX, where X represents a good leaving group, proceed as follows:

$$\underset{X}{\overset{R^1}{>}}C=O \xrightarrow{R^2Li} R^1-\underset{X}{\overset{R^2}{\underset{|}{C}}}-O^\ominus Li^\oplus \xrightarrow{-LiX} \underset{R^1}{\overset{R^2}{>}}C=O$$

$$\xrightarrow{R^2Li} R^2-\underset{R^1}{\overset{R^2}{\underset{|}{C}}}-O^\ominus Li^\oplus \xrightarrow{H_3O^+} R^2-\underset{R^1}{\overset{R^2}{\underset{|}{C}}}-OH$$

Such reactions readily give good yields of tertiary alcohols (*1, 2, 3*) (secondary alcohols from formates); the only serious limitation is the occurrence of α-deprotonation. Barbier-type reactions may also be used to synthesize tertiary alcohols (*4*). However, it is difficult to achieve good yields of ketones from these reactions, since even if a deficiency of the organolithium compound is employed both the ketone and the organolithium compound are liable to be present in the reaction mixture simultaneously. For example, even when 2-lithio-1,3-dithiane was added to a tenfold excess of ethyl benzoate at $-20°$, the main product, 2-benzoyl-1,3-dithiane, was contaminated by bis(1,3-dithian-2-yl)phenylmethanol (*5*). Nevertheless, acceptable yields of ketones are often obtainable from all these types of compound under appropriate conditions. The best results are obtained when the organolithium compound is added to an excess of the acyl derivative at a low temperature. Some examples are listed in Table 6.2, and a typical reaction with an anhydride is described below. Table 6.2 also includes examples of some related reactions. With formate esters the product is an aldehyde, and an example of this type of reaction is also described fully:

$$R^1Li + HCOOR^2 \longrightarrow R^1CH(OLi)OR^2 \xrightarrow{-LiOR^2} R^1CHO$$

Similarly, a reaction with a chloroformate can give an ester:

$$R^1Li + ClCOOR^2 \longrightarrow \left[R^1-\underset{Cl}{\overset{O^\ominus Li^\oplus}{\underset{|}{C}}}-OR^2 \right] \longrightarrow R^1COOR^2$$

TABLE 6.2

Syntheses of Carbonyl Compounds by Reactions of Organolithium Compounds with Acyl Derivatives

Acyl derivative	Organolithium compound	Product (yield %)	Ref.
MeCOCl	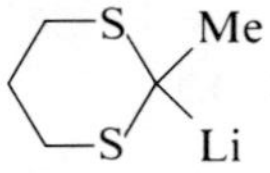 2-methyl-1,3-dithian-2-yl lithium (Me, Li on C between two S of 1,3-dithiane)	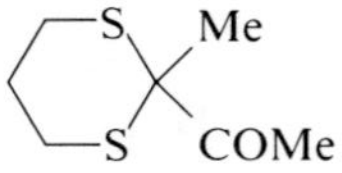 2-methyl-2-acetyl-1,3-dithiane (Me, COMe on C between two S) (50)	(2)
Me$_2$CHCOCl	LiO$_2$CCH(Li)COOEt	Me$_2$CHCOCH$_2$COOEt[a] (80)	(6)
PhCOCl	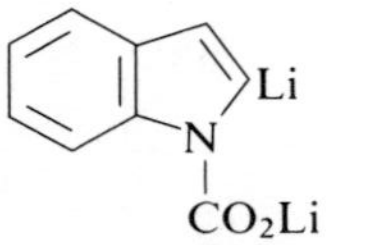 2-lithio-1-(lithiocarboxy)indole (indole with Li at C-2 and CO$_2$Li on N)	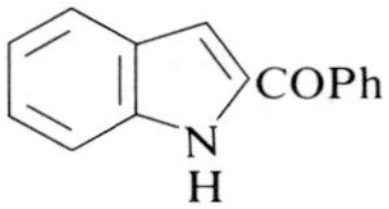 2-benzoylindole (indole with COPh at C-2, NH) (59)[a]	(7)
ClCOOMe	ClCH$_2$C≡CLi	ClCH$_2$C≡CCOOMe (81–83)	(7a)
ClCOOEt	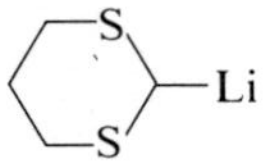 2-lithio-1,3-dithiane	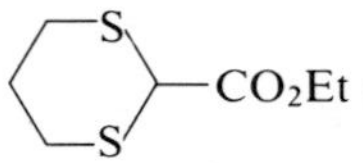 2-(ethoxycarbonyl)-1,3-dithiane (76)	(2)
ClCOOEt	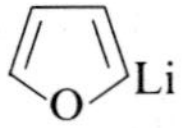 2-lithiofuran	 2-(ethoxycarbonyl)furan (61)	(8)
ClCOOEt	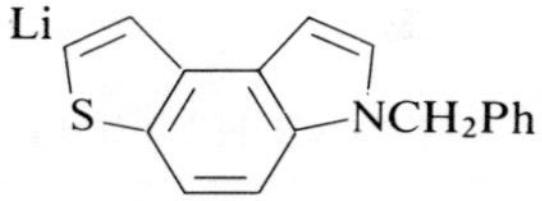 4-(benzylamino)-2-lithiothieno-fused indole system (Li on thiophene C, NCH$_2$Ph on pyrrole N)	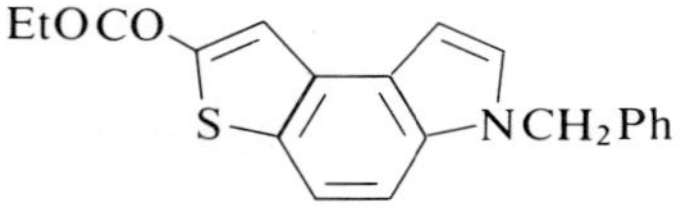 corresponding 2-ethoxycarbonyl derivative (EtOCO on thiophene C, NCH$_2$Ph on N) (96)	(9)

TABLE 6.2—continued

Electrophile	Organolithium	Product (Yield %)	Ref.
(MeCO)₂O	(EtO)₃CCH₂C≡CLi	(EtO)₃CCH₂C≡CCOMe (47)	(10)
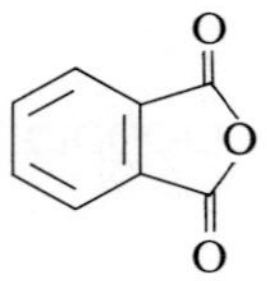 ![phthalic anhydride]	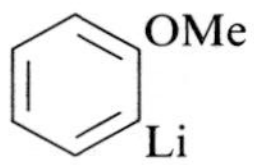 ![o-methoxyphenyllithium] (OMe, Li)	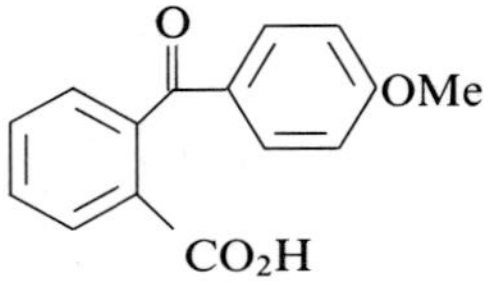 2-(4-methoxybenzoyl)benzoic acid (70)	(11)
HCOOEt	PhP(S)(Me)CH₂Li	PhP(S)(Me)CH₂CHO (42)	(12)
HCOOEt	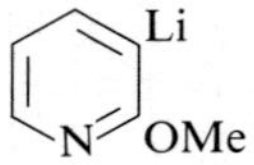 3-Li-2-OMe-pyridine	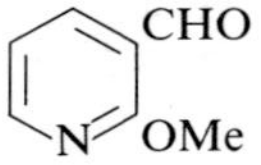 3-CHO-2-OMe-pyridine (85)	(13)
PhCOOEt	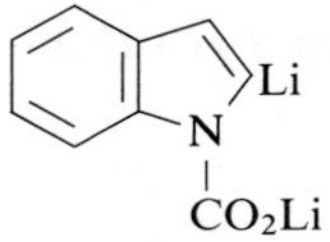 1-CO₂Li-2-Li-indole	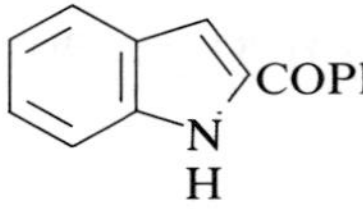 2-COPh-indole[a] (52)	(7)
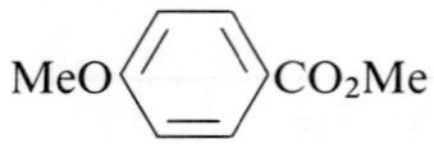 MeO–C₆H₄–CO₂Me	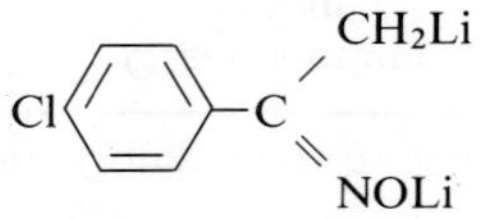 4-Cl-C₆H₄-C(CH₂Li)=NOLi	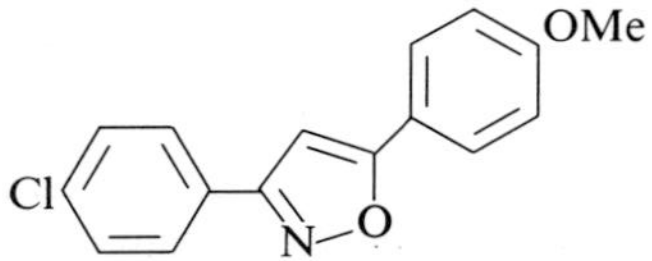 3-(4-ClC₆H₄)-5-(4-MeOC₆H₄)isoxazole (52–53)	(14)

TABLE 6.2—continued

Substrate	Organolithium	Product (yield %)	Ref.
CF$_3$COOEt	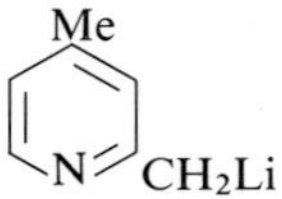 4-Me-pyridine-2-CH$_2$Li	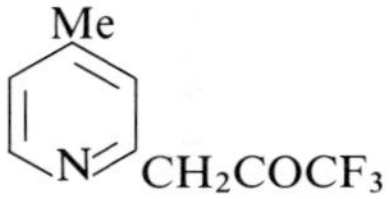 4-Me-pyridine-2-CH$_2$COCF$_3$ (91)[b]	(15)
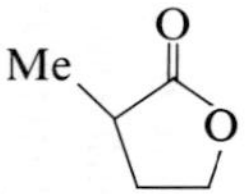 3-Me-γ-butyrolactone	BuLi	BuCOCH(Me)CH$_2$CH$_2$OH (85)	(16)
(EtO)$_2$CO	PhP(S)(Me)(CH$_2$Li)	PhP(S)(Me)(CH$_2$COOEt) (40)	(12)
(MeO)$_2$CO	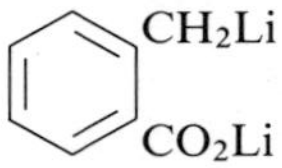 1,2-C$_6$H$_4$(CH$_2$Li)(CO$_2$Li)	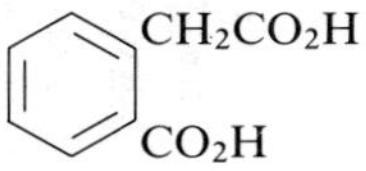 1,2-C$_6$H$_4$(CH$_2$CO$_2$H)(CO$_2$H) (85)[c]	(17)
(EtO)$_2$CO	2Bu^{t}Li	Bu$^t_2$CO (70)	(18)

[a] Following hydrolysis and decarboxylation *in situ*. [b] A good description of the preparation and addition of the organolithium compound is given but the reaction conditions for this special case would be unsatisfactory in general. [c] Ester hydrolysed on work-up.

Reactions with dialkyl carbonates are also most commonly used to prepare esters, but where the product is sterically hindered it may be possible to obtain a symmetrical ketone (see final entry in Table 6.2).

In cases where alcohol formation occurs even under the most favourable conditions, some chemical modifications may be employed. For reactions with acyl halides, the earlier use of cadmium halides as additives has been largely superseded by the conversion of the organolithium compound into an organocopper reagent *in situ* (*19, 20*). Organomanganese reagents, though less fully explored, may also be used (*21, 22*). There is no analogous general modification for reactions with esters, but in some cases the presence of trimethylsilyl chloride in the reaction mixture is beneficial; the silyl chloride reacts faster with the adduct than it does with the organolithium compound (*23*):

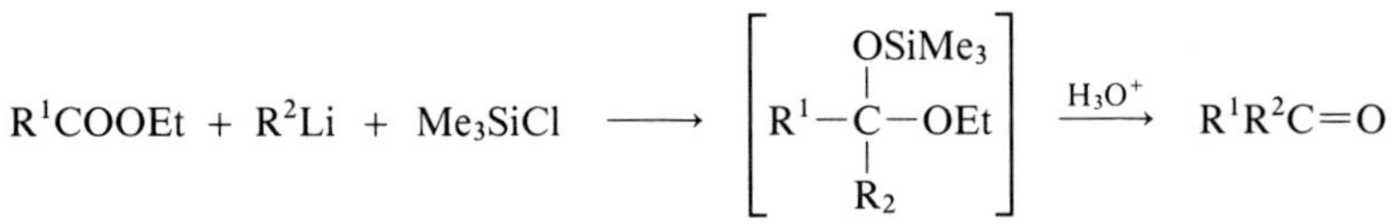

$$R^1COOEt + R^2Li + Me_3SiCl \longrightarrow \left[R^1-\underset{R_2}{\underset{|}{\overset{OSiMe_3}{\overset{|}{C}}}}-OEt \right] \xrightarrow{H_3O^+} R^1R^2C=O$$

For αβ-unsaturated esters the situation is much the same as for αβ-unsaturated ketones. Organolithium compounds tend to add to the carbonyl group (see Ref. (*2*) for an example); 1,2-addition under Barbier conditions has also been described (*24*). Conjugate addition occurs with organocopper reagents, though the lower reactivity of esters compared to ketones can be a limitation (*25*). Conjugate addition may also be induced by steric hindrance, as in the following example (*26*):

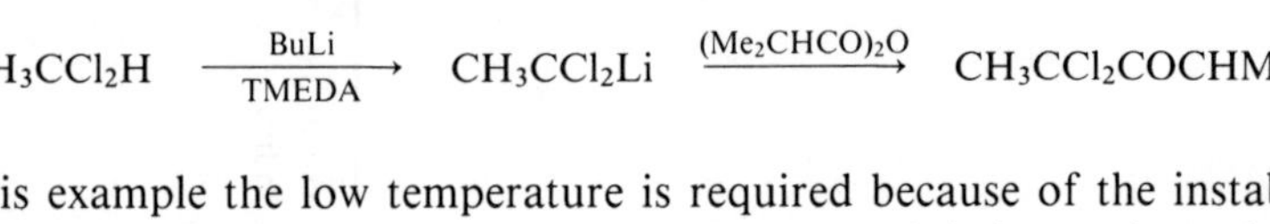

Reaction with an anhydride; 2,2-dichloro-4-methylpentan-3-one (27)

$$CH_3CCl_2H \xrightarrow[TMEDA]{BuLi} CH_3CCl_2Li \xrightarrow{(Me_2CHCO)_2O} CH_3CCl_2COCHMe_2$$

In this example the low temperature is required because of the instability of the organolithium intermediate as well as to minimize tertiary alcohol formation.

TMEDA (4.8 g, 41 mmol) is dissolved in a mixture of ether (80 ml) and THF (40 ml) in a 250 ml three-necked flask fitted with a septum, a pressure-equalizing dropping funnel, a gas inlet and a stirrer, and flushed with

6.2. REACTIONS WITH ACYL HALIDES ETC.

nitrogen. The solution is cooled and stirred well as n-butyllithium (in hexane, 40 mmol) is added. The mixture is cooled to $-100°$ and stirred well as a solution of 1,1-dichloroethane (5.0 g, 50 mmol) in THF (20 ml) is added dropwise, slowly. The mixture is stirred at $—100°$ for 1 h, during which time an abundant white precipitate forms. The well-stirred suspension is maintained at $-95°$ as isobutyric anhydride (2-methylpropanoic anhydride) is added, and stirring is continued for 2 h. Water and pentane are added and the mixture is allowed to warm to room temperature. The organic layer is separated, dried (MgSO$_4$), and the solvent is evaporated. To the residue is added methanol (20 ml) containing a few drops of sulphuric acid. The methanol is evaporated, and saturated aqueous sodium hydrogencarbonate is added to the residue until effervescence ceases. The mixture is extracted three times with pentane. The combined extracts are dried (MgSO$_4$) and evaporated. Distillation of the residue gives 2,2-dichloro-4-methylpentan-3-one (86%), b.p. 143–145°.

Reaction with methyl formate; 2,2-dichloro-2-phenylethanal (28)

$$PhCCl_3 \xrightarrow{Bu^sLi} PhCCl_2Li \xrightarrow[\text{(ii) } H_3O^+]{\text{(i) HCOOMe}} PhCCl_2CH(OH)OMe$$

$$\xrightarrow{\Delta} PhCCl_2CHO$$

In this case, unusually, the hemiacetal initially formed is comparatively stable, but the free aldehyde is obtained on distillation. The temperature used for the reaction is critical, and must be carefully controlled.

ααα-Trichlorotoluene (benzotrichloride) (4.89 g, 25 mmol), THF (60 ml), ether (40 ml) and pentane (40 ml) are placed in a four-necked flask equipped with a magnetic stirrer, a low-temperature (pentane) thermometer, a pressure-equalizing dropping funnel, a septum and a gas inlet tube. An atmosphere of nitrogen is maintained as the solution is cooled to $-90°$. The solution is maintained at $-90°$ or slightly below as s-butyllithium (ca. 1.5 M in hexane or cyclohexane, 25 mmol) is added dropwise with stirring during 20 min, via the septum. When a precipitate appears, the mixture is cooled to $-110°$, and maintained at that temperature as a solution of methyl formate (1.8 g, ca. 30 mmol) in ether (10 ml) is added dropwise with stirring during 10 min. The temperature is allowed to rise to $-90°$. M Sulphuric acid (50 ml) is added rapidly to the stirred mixture. The organic layer is separated, and the aqueous layer is extracted with pentane (3 × 50 ml). The combined aqueous layers are washed with saturated aqueous sodium chloride and dried (MgSO$_4$). The solvents are evaporated[a] and the residue is distilled under reduced pressure to give 2,2-dichloro-2-phenylethanal (70–80%), b.p. 56–58°/0.8 mmHg, 104°/14 mmHg.

[a] An NMR spectrum of the crude product, before distillation, showed only a weak signal for the formyl proton at δ 9.2, but also a signal at δ ca. 3.6 assigned to a methoxy group (29).

References

1. See Section 6.1.1, Ref. (2).
2. D. Seebach and E. J. Corey, *J. Org. Chem.* **40**, 231 (1975).
3. H. E. Zimmerman and J. M. Nuss, *J. Org. Chem.* **51**, 4604 (1986).
4. See Section 6.1.1, Ref. (13).
5. E Juaristi, J. Tapia and R. Mendez, *Tetrahedron* **42**, 1253 (1986).
6. W. Wieringa and H. I. Skulnik, *Org. Synth.* **61**, 5 (1983).
7. A. R. Katritzky and K. Akutagawa, *Tetrahedron Lett.* **26**, 5935 (1985).
7a. M. Olomucki and J. Y. Le Gall, *Org. Synth.* **65**, 47 (1987).
8. F. Pinkerton and S. F. Thames, *J. Heterocycl. Chem.* **7**, 747 (1970).
9. V. H. Rawal, R. J. Jones and M. P. Cava, *J. Org. Chem.* **52**, 19 (1987).
10. R. Finding and U. Schmidt, *Angew. Chem. Int. Ed. Engl.* **9**, 456 (1970).
11. W. E. Parham and R. M. Piccirilli, *J. Org. Chem.* **41**, 1268 (1968).
12. F. Mathey and F. Mercier, *J. Organomet. Chem.* **177**, 255 (1979).
13. F. Trécourt, J. Morel and G. Quéguiner, *J. Chem. Res. (S)* 46, (M) 536 (1979).
14. M. Perkins and C. F. Beam, *Org. Synth.* **55**, 39 (1976).
15. R. Levine, D. A. Dimmig and W. M. Kadunce, *J. Org. Chem.* **39**, 3834 (1974); see also X. Creary, *J. Org. Chem.* **52**, 5026 (1987).
16. S. Cavicchioli, D. Savoia, C. Trombini and A. Umani-Ronchi, *J. Org. Chem.* **49**, 1246 (1984).
17. F. M. Hauser and R. Rhee, *Synthesis* 245 (1977).
18. J.-E. Dubois, B Leheup, F. Hennequin and P. Bauer, *Bull. Soc. Chim. Fr.* 1150 (1967).
19. G. H. Posner, *Org. React.* **22**, 253 (1975).
20. See Section 6.1.2, Ref. (3b).
21. G. Friour, G. Cahiez and J. Normant, *Synthesis* 37 (1984) and 50 (1985).
22. G. Cahiez, J. Rivas-Enterrios and H. Granger-Veyron, *Tetrahedron Lett.* **27**, 4441 (1986).
23. M. P. Cooke, *J. Org. Chem.* **51**, 951 (1986).
24. P. J. Pearce, D. H. Richards and N. F. Scilly, *Org. Synth* **52**, 19 (1972).
25. See Section 6.1.2, Ref. (3).
26. M. P. Cooke, *J. Org. Chem.* **51**, 1637 (1986); see also M. P. Cooke, *J. Org. Chem.* **49**, 1146 (1984).
27. J. Villieras, P. Perriot and J. F. Normant, *Bull. Soc. Chim. Fr.* 765 (1977).
28. J. Villieras and M. Rambaud, *Synthesis* 644 (1980).
29. C. L. Cheong and B. J. Wakefield, Unpublished work.

6.3. ADDITION TO N,N-DISUBSTITUTED AMIDES

The adducts of organolithium compounds with N,N-disubstituted amides are comparatively stable, so that it is usually possible to avoid the elimination and subsequent alcohol formation shown on p. 76. However, the widely used synthesis of aldehydes from N,N-dialkylformamides and organolithium

6.3. ADDITION TO N,N-DISUBSTITUTED AMIDES

compounds is somewhat unpredictable. Sometimes the reaction with N,N-dimethylformamide (DMF) works well (see Table 6.3) but in other cases it gives poor yields, or even fails completely (1). N-Methylformanilide (see General Ref. A and Ref. (2)), Comins' reagent [1] (3) and N-formylpiperidine (4) are possible alternatives. A Barbier-type procedure, assisted by ultrasonic irradiation, has been reported to give good results (5).

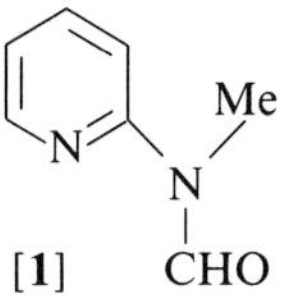

[structure of compound [1]: 2-pyridyl-N(Me)-CHO]

The examples of syntheses of aldehydes by reactions of organolithium compounds with N,N-dialkylformamides listed in Table 6.3 are recent; older examples are tabulated in General Ref. A (p. 141), and a detailed procedure is described below.

Reactions of organolithium compounds with other N,N-disubstituted amides have been less commonly employed, but can give good yields of ketones; examples are listed in Table 6.4 and others are tabulated in General Ref. D(ii) (p. 35).* The following is an example of an intramolecular reaction of this type (12):

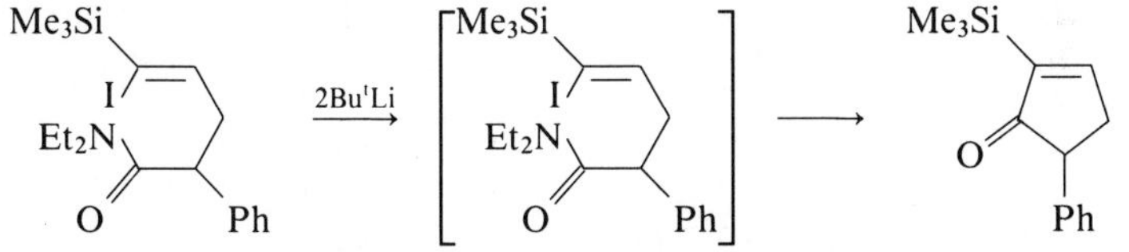

[reaction scheme showing Me$_3$Si and Et$_2$N substituted cyclohexanone with Ph, treated with 2Bu^{t}Li, giving cyclopentenone product]

Table 6.4 also includes examples of reactions with chlorocarbamates, giving amides and with a carbamate ester, giving a symmetrical ketone:

$$R^1Li + ClCONR_2^2 \longrightarrow R^1CONR_2^2$$

$$R^1Li + R^2OCONR_2^3 \longrightarrow R^1CONR_2^3 \xrightarrow{R^1Li} R_2^1CO$$

The adduct of an organolithium compound and a dialkylamide is in effect a protected carbonyl compound, and in suitable cases further "one-pot"

---

* It should be noted that aromatic amides may be sufficiently unreactive towards organolithium compounds for *ortho*-metallation to occur rather than addition to the carbonyl group (see Section 3.2).

TABLE 6.3

Synthesis of Aldehydes by Reactions of Organolithium Compounds with N,N-Dialkylformamides

Formamide	Organolithium compound	Product (yield %)	Ref.
DMF	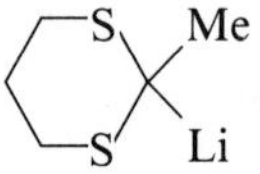 2-methyl-2-lithio-1,3-dithiane	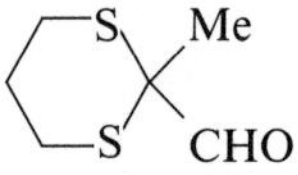 2-methyl-2-formyl-1,3-dithiane (>57)	(6)
DMF	PhCCl$_2$Li	PhCCl$_2$CHO (68)[a]	(7)
DMF	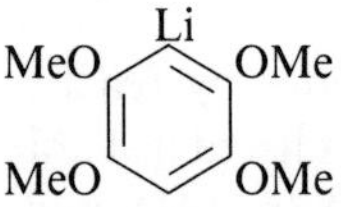 2,3,5,6-tetramethoxyphenyllithium	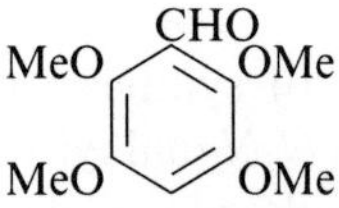 2,3,5,6-tetramethoxybenzaldehyde (72)	(8)
DMF	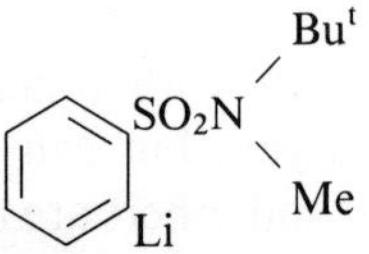 2-[N-t-butyl-N-methylsulfamoyl]phenyllithium	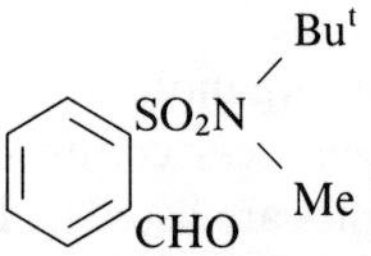 2-[N-t-butyl-N-methylsulfamoyl]benzaldehyde (81)	(9)
DMF	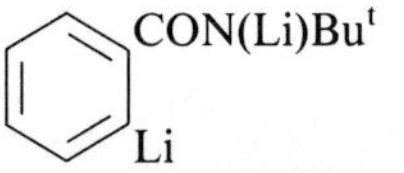 2-[N-lithio-N-t-butylcarbamoyl]phenyllithium	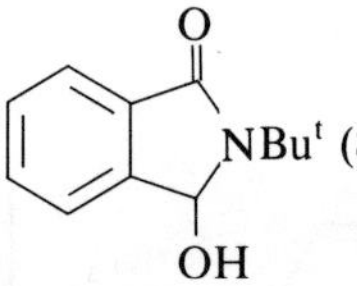 3-hydroxy-2-t-butyl-2,3-dihydro-1H-isoindol-1-one (82)[b]	(9)
DMF	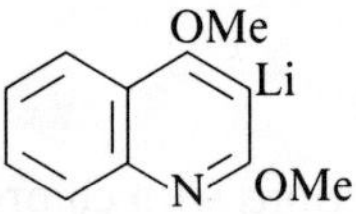 2,4-dimethoxy-3-lithioquinoline	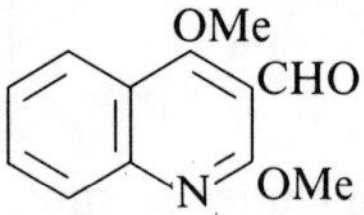 2,4-dimethoxyquinoline-3-carbaldehyde (94)	(10)
PhN(Me)CHO	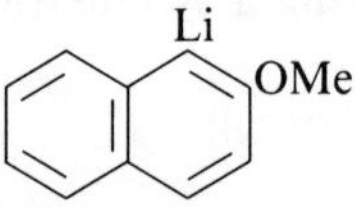 3-methoxy-2-lithionaphthalene	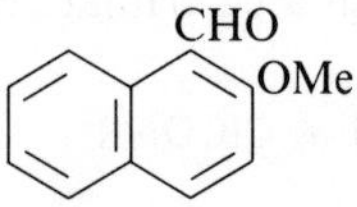 3-methoxynaphthalene-2-carbaldehyde (77)	(2)
[1]	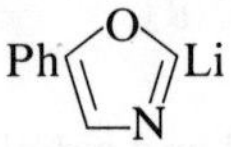 2-lithio-5-phenyloxazole	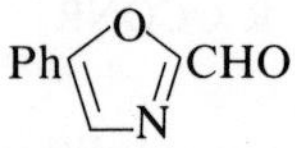 5-phenyloxazole-2-carbaldehyde (61)	(11)
 cyclohexyl-CHO	PhLi	PhCHO (94)	(4)

[a] The corresponding reaction with methyl formate, described in Section 6.2, gives better results in this case. [b] Cyclized product obtained after acid work-up.

6.3. ADDITION TO N,N-DISUBSTITUTED AMIDES

reactions can be carried out. One of the first examples is shown below (17);*
for other examples see Refs (19) and (20):†

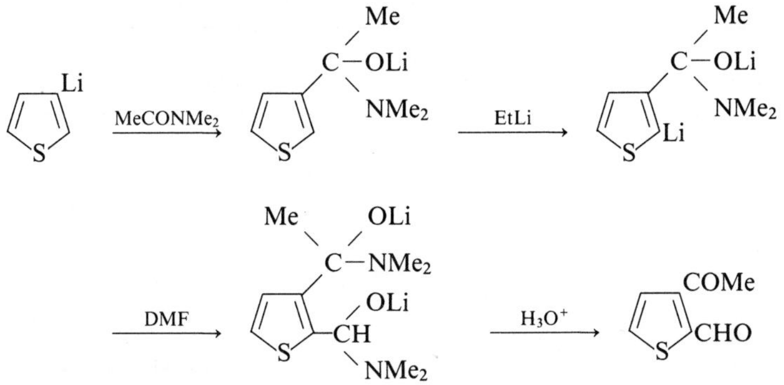

[Scheme showing thiophene-Li reacting with MeCONMe₂ to give 2-thienyl-C(Me)(OLi)NMe₂, then with EtLi to give the 5-lithiated species, then with DMF to give a thiophene bearing both the C(Me)(OLi)NMe₂ group and a CH(OLi)NMe₂ group, then H₃O⁺ to give 2-acetyl-5-formylthiophene (COMe and CHO).]

$\alpha\beta$-Unsaturated amides, both secondary and tertiary, are more prone to conjugate addition than $\alpha\beta$-unsaturated esters and ketones; with the amides this mode of addition has been found to occur even without special steric or solvent effects or copper catalysis (22, 23). Such reactions are the basis for useful tandem addition–alkylation sequences:

$$R^1CH=CHCONR^2R^3 \xrightarrow{R^4Li} R^1R^4CHCH=CNR^2R^3$$
$$\phantom{R^1CH=CHCONR^2R^3 \xrightarrow{R^4Li} R^1R^4CHCH}\underset{O}{\overset{\ominus}{|}}$$

$$\xrightarrow{E^+} R^1R^4CHCHCONR^2R^3$$
$$|$$
$$E$$

It should be noted, however, that "anti-Michael" additions may occur as well as (or even in place of) Michael addition to $\alpha\beta$-unsaturated secondary amides, particularly when the group X is carbanion-stabilizing (24):

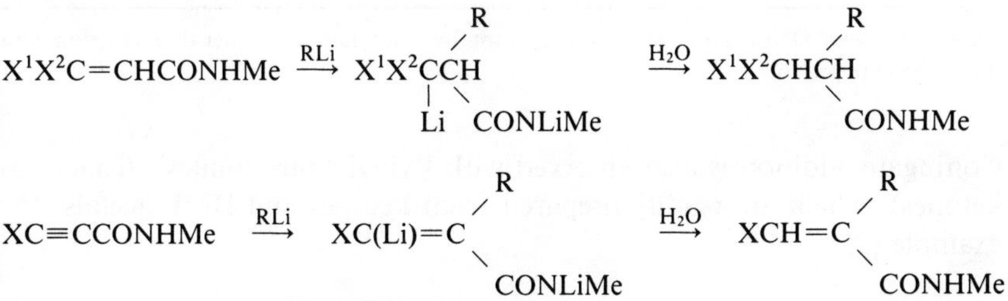

$$X^1X^2C=CHCONHMe \xrightarrow{RLi} X^1X^2C\overset{R}{\underset{\underset{CONLiMe}{|}}{C}}H \xrightarrow{H_2O} X^1X^2CH\overset{R}{\underset{CONHMe}{C}}H$$

$$XC{\equiv}CCONHMe \xrightarrow{RLi} XC(Li)=C\overset{R}{\underset{CONLiMe}{\diagdown}} \xrightarrow{H_2O} XCH=C\overset{R}{\underset{CONHMe}{\diagdown}}$$

---

* The paper does not give experimental details for this example, but it is a particularly apposite one for this section. Sequences involving reactions of lithiated DMF adducts with tributyl borate and with carbon dioxide are fully described (17).

† Similar intermediates can also be prepared by reactions of lithium dialkylamides with aromatic aldehydes (21).

TABLE 6.4

Reactions of Organolithium Compounds with *N*,*N*-Dialkyl Amides, Chlorocarbamates and Carbamate Esters

Amide etc.	Organolithium compound	Product (yield %)	Ref.
 MeCON(Me)(Me)	PhLi	PhCOMe (77)	(13)
Me$_2$CHCH$_2$CONMe$_2$	CH$_2$=C(CH(OEt)$_2$)(Li)	CH$_2$=C(CH(OEt)$_2$)(COCH$_2$CHMe$_2$) (70)	(14)
PhCONMe$_2$	MeLi	PhCOMe (83)	(1)
ClCONEt$_2$	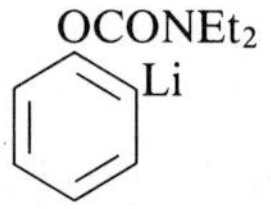 o-Li-C$_6$H$_4$-OCONEt$_2$ (*a*)	 o-(CONEt$_2$)-C$_6$H$_4$-OCONEt$_2$ (89)	(15)
ClCONEt$_2$	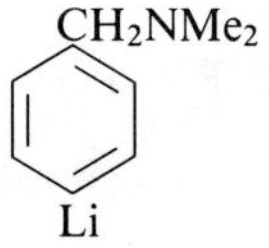 o-Li-C$_6$H$_4$-CH$_2$NMe$_2$	 o-(CONEt$_2$)-C$_6$H$_4$-CH$_2$NMe (65)	(16)
EtOCONMe$_2$	 2-thienyl-Li	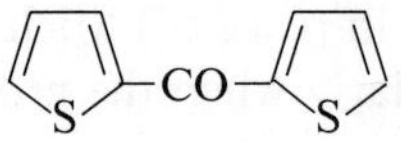 (2-thienyl)CO(2-thienyl) (92)	(17)

a Note that PhOCONEt$_2$ undergoes *O*-metallation by s-butyllithium rather than reaction with the carbamate group.

Conjugate addition is also observed with "vinylogous amides" (enaminoketones), which are readily prepared from ketones and DMF acetals; for example (25)

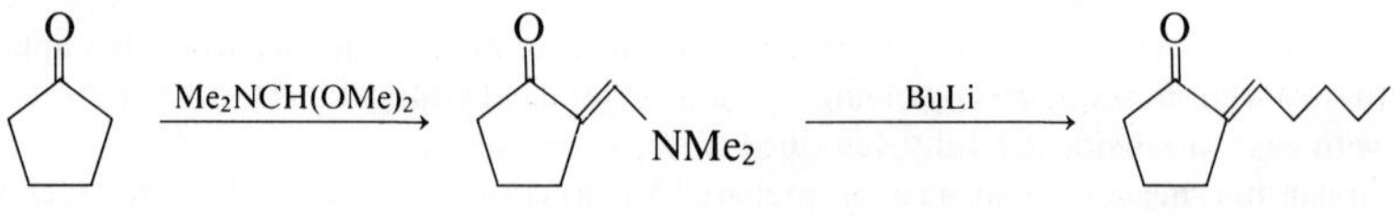

cyclopentanone $\xrightarrow{\text{Me}_2\text{NCH(OMe)}_2}$ 2-(dimethylaminomethylene)cyclopentanone $\xrightarrow{\text{BuLi}}$ 2-pentylidenecyclopentanone

75%

6.3. ADDITION TO N,N-DISUBSTITUTED AMIDES

Reaction with DMF: 3-chloro-2-(methoxymethoxy)benzaldehyde (26)

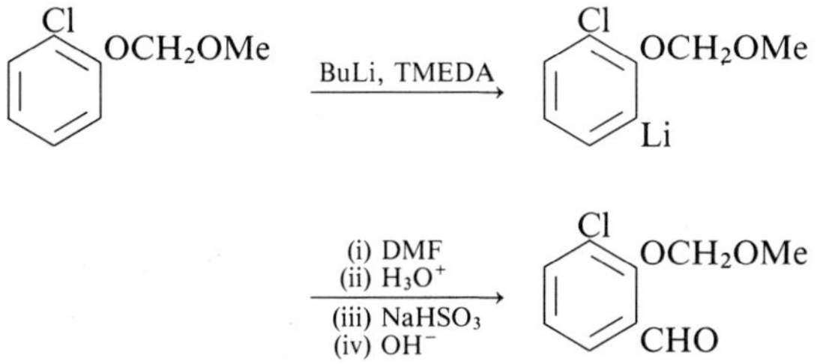

TMEDA (53.0 g, 0.5 mmol) is added to n-butyllithium (*ca.* 2 M in hexane, 0.5 mol). The solution is cooled to 0° and stirred as 1-chloro-2-(methoxymethoxy)benzene (86.3 g, 0.5 mol) is added at 0–5° during 30 min. Stirring at this temperature is continued for 30 min. The resulting yellow slurry is transferred under nitrogen to a pressure-equalizing dropping funnel attached to a 2 l three-necked flask containing a solution of DMF (43.8 g, 0.6 mol) in xylene (470 ml). The DMF solution is maintained at 0–5° and stirred vigorously as the organolithium slurry is added during 25 min. Stirring is continued for 1 h. The reaction mixture is then transferred slowly through a polyethylene tube into a stirred mixture of concentrated hydrochloric acid (190 ml) and crushed ice (900 ml), the temperature being kept below 5°. When the transfer is complete, stirring is continued for 20 min. The organic phase is separated and washed successively with ice-cold M hydrochloric acid (200 ml) and saturated aqueous sodium chloride (200 ml). The organic layer is then stirred for 20 min with a solution of sodium bisulphite (52 g, 0.5 mol) in water (120 ml) mixed with ice (150 g). The aqueous layer is separated and kept cold. The extraction of the organic layer is repeated twice, but each time with half the original quantities. The bisulphite extracts are combined and kept below 10° as an aqueous solution of sodium hydroxide (45 g) is added slowly to the stirred mixture, the pH being finally adjusted to 11. The resulting crystalline precipitate is collected by suction filtration, washed with water, and air dried to give 3-chloro-2-(methoxymethoxy)benzaldehyde (85.4 g, 85%), m.p. 38–40°.

References

1. G. A. Olah, G. K. S. Prakash and M. Arvanaghi, *Synthesis* **228** (1984).
2. N. S. Narasimhan and R. S. Mali, *Tetrahedron* **31**, 1005 (1975).
3. D. Comins and A. I. Meyers, *Synthesis* 403 (1978).
4. G. A. Olah and M. Arvanaghi, *Agnew. Chem. Int. Ed. Engl.* **20**, 878 (1981).

5. C. Pétrier, A. L. Gemal and J.-L. Luche, *Tetrahedron Lett.* **23**, 3361 (1982); J. Einhorn and J. L. Luche, *Ibid.* **27**, 1791 (1986).
6. See Section 6.2, Ref. (*2*).
7. See Section 6.6, Ref. (*28*).
8. F. Dallacker and G. Sanders, *Chem.-Ztg* **110**, 369 (1986).
9. See Section 3.2, Ref. (*3*).
10. N. S. Narasimhan and S. P. Bhagwat, *Synthesis* 903 (1979).
11. L. N. Pridgen and S. C. Shilcrat, *Synthesis* 1048 (1984).
12. H. Sawada, M. Webb, A. T. Stoll and E.-I. Negishi, *Tetrahedron Lett.* **27**, 775 (1986).
13. S. Wattanasin and F. G. Kathawala, *Tetrahedron Lett.* **25**, 811 (1984).
14. J. C. Depezay, Y. Le Merrer and M. Sanière, *Synthesis* 766 (1985).
15. M. P. Sibi, S. Chattopadhyay, J. W. Dankwards and V. Snieckus, *J. Am. Chem. Soc.* **107**, 6312 (1985).
16. See Section 3.2, Ref. (*37*).
17. U. Michael and A.-B. Hörnfeldt, *Tetrahedron Lett.* 5219 (1970).
18. U. Michael and S. Gronowitz, *Acta Chem. Scand.* **22**, 1353 (1968).
19. L. Barsky, H. W. Gschwend, J. McKenna and H. R. Rodriguez, *J. Org. Chem.* **41**, 3651 (1976).
20. J. Einhorn and J. L. Luche, *Tetrahedron Lett.* **27**, 1791 (1986).
21. D. L. Comins, J. D. Brown and N. B. Mantlo, *Tetrahedron Lett.* **23**, 3979 (1982); D. L. Comins and J. D. Brown, *Ibid.* **24**, 5465 (1983); D. L. Comins and J. D. Brown, *J. Org. Chem.* **49**, 1078 (1984); D. L. Comins and M. O. Killpack, *Ibid.* **52**, 104 (1987).
22. J. E. Baldwin and W. A. Dupont, *Tetrahedron Lett.* **21**, 1881 (1980).
23. G. P. Mpango, K. K. Mahalanabis, Z. Mahdavi-Damghani and V. Snieckus, *Tetrahedron Lett.* **21**, 4823 (1980); G. P. Mpango and V. Snieckus, *Ibid.* **21**, 4827 (1980).
24. G. W. Klumpp, A. J. C. Mierop, J. J. Vrielink, A. Brugman and M. Schakel, *J. Am. Chem. Soc.* **107**, 6740 (1985).
25. R. F. Abdulla and K. H. Fuhr, *J. Org. Chem.* **43**, 4248 (1978).
26. See Section 3.2, Ref. (*53*).

6.4. ADDITION TO CUMULATED CARBONYL GROUPS AND CARBOXYLATE IONS

Reactions of organolithium compounds with cumulated carbonyl groups, apart from the familiar ones with carbon dioxide, have been under-exploited. The reaction with ketenes is a promising method for preparing enolates, particularly as they are formed stereospecifically in some cases, such as the following (*1*):

6.4. ADDITION TO CUMULATED –CO– GROUPS AND –CO.O⁻ IONS

The reaction with isocyanates is a useful method for synthesizing amides:*

$$RLi + R^2N=C=O \longrightarrow R^2N=C\begin{matrix}R^1\\ \\OLi\end{matrix} \xrightarrow{H_3O^+} R^2NHCOR^1$$

Examples of both of these types of reaction are shown in Table 6.5.

Although the reaction of organolithium compounds with carbon dioxide, giving lithium carboxylates and thence carboxylic acids, is so familiar, it is not without its problems; it is linked here with the associated further reactions of the carboxylate to give, after hydrolysis, a ketone (and unwanted still further reaction to give a tertiary alcohol):

$$RLi + CO_2 \longrightarrow R-C\begin{matrix}O\\ \\O\end{matrix}Li^\oplus \xrightarrow{RLi} R_2C\begin{matrix}OLi\\ \\OLi\end{matrix} \xrightarrow[-Li_2O]{H_2O\ or} R_2C=O$$

$$\downarrow H_3O^+ \qquad\qquad\qquad\qquad\qquad \downarrow RLi$$

$$RCO_2H \qquad\qquad R_3COH \xleftarrow{H_3O^+} R_3COLi$$

Of these reactions, the one leading to the carboxylic acid is most commonly desired. The ones leading to ketone or alcohol are often unwanted sidereactions (though good yields of symmetrical ketones can sometimes be obtained (*8, 9*)). However, the reaction of an organolithium compound with a carboxylate salt provides a very useful synthesis of unsymmetrical ketones.

In order to obtain the best yield of a carboxylic acid, it is necessary to avoid as far as possible the simultaneous presence of the organolithium compound and the carboxylate salt, particularly at other than low temperatures. This is most simply accomplished by pouring a solution of the organolithium compound onto solid carbon dioxide. The most convenient vessel is a widemouthed conical flask or a tall beaker. If the flask is flushed with dry nitrogen and fresh *dry* solid carbon dioxide is placed in the vessel, which is then covered with a clock glass, the evaporating carbon dioxide effectively blankets the solid. The solid carbon dioxide should be crushed, and it is sometimes advantageous to cover it with a dry solvent such as ether. The organolithium solution is then added by syringe, by syphoning from the vessel in which it is prepared, or even simply by pouring. The metal apparatus

---
* In Barbier-type syntheses from aryl halide, the metal, and an isocyanate, lithium was much less satisfactory than magnesium (*2*).

TABLE 6.5

Addition of Organolithium Compounds to Ketenes and Isocyanates

Ketene or isocyanate	Organolithium compound	Further reagent	Product (yield %)	Ref.
$Bu^t_2C=C=O$	Bu^tLi	Me_3SiCl	$Bu^t_2C=C(OSiMe_3)(Bu^t)$ (70)	(3)
		H_2O	$Bu^t_2CHCOBu^t$ (73)	(3)
$Ph_2C=C=O$	PhLi	PhCOCl	$Ph_2C=C(OCOPh)(Ph)$ (84)	(4)
$(Me_3Ge)_2C=C=O$	MeLi	H_2O	$(Me_3Ge)_2CHCOMe$ (62)	(5)
$Bu^tN=C=O$	 PhS–C(Li)=CH–NMe$_2$	Na_2CO_3 aq.	 PhS–C(NMe$_2$)=CH–CONHBu^t (a) (79)	(6)
$PhN=C=O$	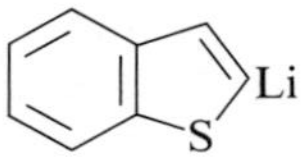 benzothiophene-Li	NH_4Cl aq.	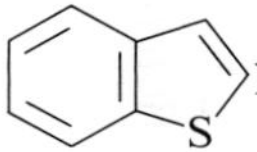 benzothiophene-NHCOPh (81)	(7)
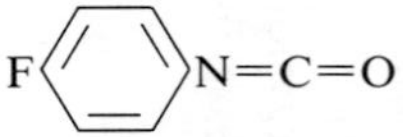 $F-C_6H_4-N=C=O$	 PhS–C(Li)=CH–NMe$_2$	Na_2CO_3 aq.	 PhS–C(NMe$_2$)=CH–CONHC$_6$H$_4$F (47)	(6)

[a] Isolated as hydrochloride

TABLE 6.6

Synthesis of Carboxylic Acids by Reaction of Organolithium Compounds with Carbon Dioxide

Organolithium compound	Yield of corresponding carboxylic acid (%)	Ref.
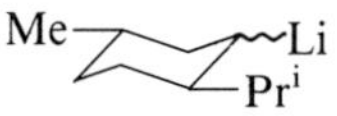 Me-cyclohexyl-Li, Pr^i (a)	>90	(11)
 norbornenyl-Li (b)	64	(12)
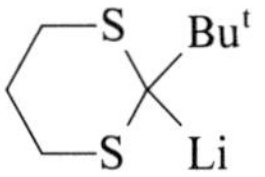 dithiane with Bu^t, Li	76	(13)
$H_2C=C=CHLi$	80^c	(14)
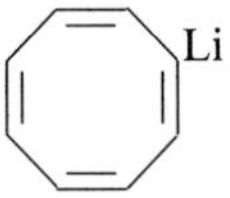 cyclooctatetraenyl-Li	59	(15)
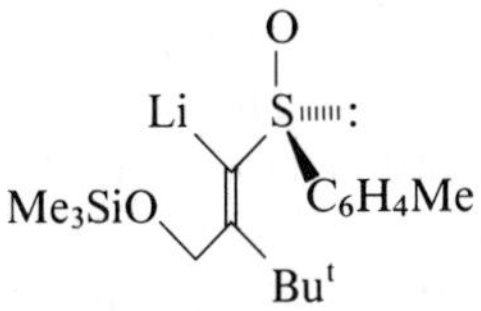 Me$_3$SiO, Li, S(=O)-C$_6$H$_4$Me, Bu^t	(78)^{d,e}	(16)
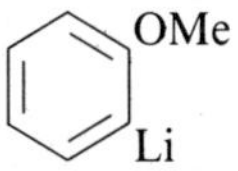 OMe-phenyl-Li	65	(17)
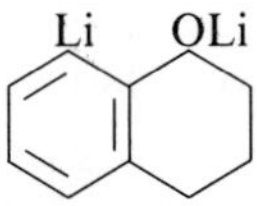 Li, OLi tetralin	63^f	(18)
C_6Cl_5Li	66^g	(19)

^{a} Carboxylation occurs with retention of configuration. ^{b} See Section 3.1.2. ^{c} A solution of carbon dioxide in THF at −90° was used. ^{d} Carbon dioxide bubbled into solution at −78°. ^{e} Yield of lactone [2] obtained after treatment of reaction mixture with p-toluenesulphonic acid. ^{f} Carbon dioxide bubbled into suspension at −30°. ^{g} Solid added to solid carbon dioxide–ether slurry.

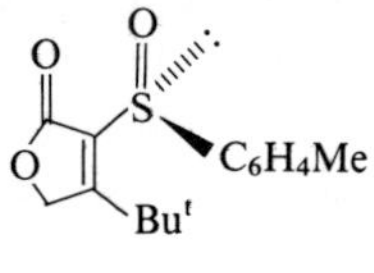

[2] furanone with S(=O)$_2$-C$_6$H$_4$Me, Bu^t

for reactions of Grignard reagents with $^{13}CO_2$, described in Ref. (10), could presumably also be used for organolithium compounds.

Some examples of syntheses of carboxylic acids from organolithium compounds are listed in Table 6.6; some of these examples illustrate alternatives to the procedure described above. The following description is of the commonly used procedure.

Reaction with carbon dioxide; 2-chloro-3,3-diphenylacrylic acid (20)

$$Ph_2C=CHCl \xrightarrow{BuLi} Ph_2C=C\underset{Cl}{\overset{Li}{\diagup}} \xrightarrow[\text{(ii) } H_3O^+]{\text{(i) } CO_2} Ph_2C=C\underset{Cl}{\overset{CO_2H}{\diagup}}$$

A solution of 1-chloro-2,2-diphenylethene (6.45 g, 30 mmol) in THF (50 ml) under nitrogen is cooled to $-71 \pm 1°$[a] and stirred at that temperature as n-butyllithium (ca. 1.3 M in diethyl ether, 30 mmol) is added during 1 h. The colour of the reaction mixture is initially pink, then yellow, and finally lightbrown. The mixture is poured onto powdered solid carbon dioxide covered with dry ether. A colourless precipitate is formed. Water is added, and the ether–THF mixture is evaporated on a rotary evaporator. The aqueous solution is extracted with ether. The aqueous layer is separated, acidified with an excess of dilute sulphuric acid, and thoroughly extracted with ether. The combined ether extracts from the acidified solution are dried ($CaCl_2$), filtered and evaporated, to yield 2-chloro-3,3-diphenylacrylic acid (6.43 g, 86%), m.p. 130–133°.[b]

[a] Careful control of the temperature is needed to minimize rearrangement of the 2,2-diphenyl-1-chloroethenyllithium. In this experiment, evaporation of the original ether extract gave a mixture of the acrylic acid (ca. 0.24 g) and diphenylacetylene (ca. 0.56 g) (20).
[b] Recrystallization from hexane gives pure acid (6.20 g), m.p. 136°.

The reaction of an organolithium compound with a carboxylic acid or a carboxylate salt, to give a ketone, has been known since 1933 (8), but was regarded as very unpredictable. However, Bare and House showed that, provided that the experimental conditions are carefully controlled, consistently high yields are achievable (21). The necessary factors have been described (9, 21); they may be summarized as follows.

(i) If the free acid is used, with two equivalents of the organolithium compound, efficient cooling and stirring is needed to control the exothermic reaction. Preferably, as in the example described below,

6.4. ADDITION TO CUMULATED –CO– GROUPS AND –CO.O⁻ IONS

the lithium carboxylate is pre-formed, for example by reaction with lithium hydride.*

(ii) The presence of unreacted organolithium compound at the end of the reaction should be avoided if possible. The hydrolysis procedure should ensure that the simultaneous presence of ketone and organolithium compound is minimized, for example by adding the reaction mixture to a large excess of the hydrolysing medium, with efficient stirring.

Tables of examples are given in Ref. (*9*), which also includes representative experimental procedures. Further examples are listed in Table 6.7, and an intramolecular reaction is shown below (*22*):

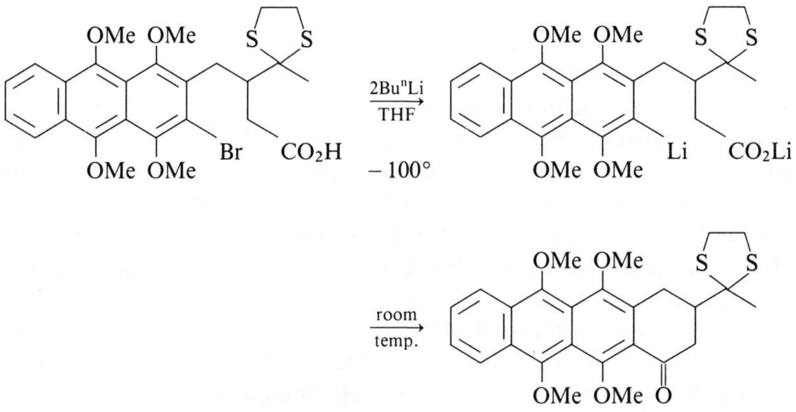

Reaction with a carboxylate; acetylcyclohexane (9, 27)

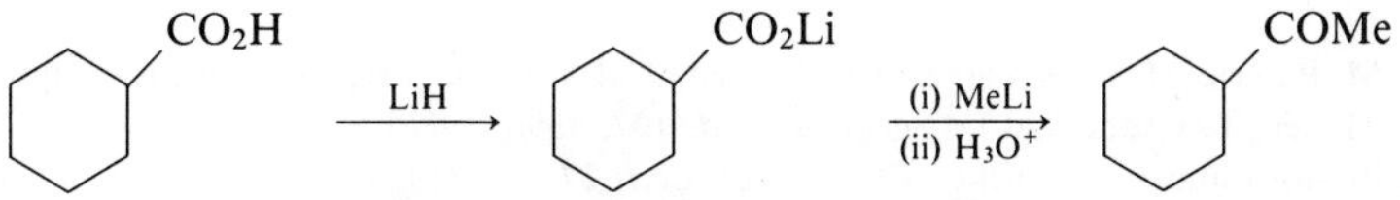

Powdered lithium hydride (1.39 g, 0.174 mol) is suspended in dry 1,2-dimethoxyethane (freshly distilled from lithium aluminium hydride, 100 ml) under nitrogen. The suspension is stirred vigorously as a solution of cyclohexanecarboxylic acid (19.25 g, 0.150 mol) in dry 1,2-dimethoxyethane (100 ml) is added. (CAUTION: vigorous evolution of hydrogen.) The resulting mixture is heated and stirred under reflux for 2.5 h. The suspension is cooled to *ca.* 10° and stirred rapidly as methyllithium (*ca.* 1.4 M in ether,

---

* Successful reactions with lithium formate, leading to aldehydes, have so far only been reported for organomagnesium compounds (*23*).

TABLE 6.7

Synthesis of Ketones by Reactions of Organolithium Compounds with Carboxylates

Carboxylate	Organolithium compound	Product (yield %)	Ref.
$CH_3CH_2CH_2CO_2Li$	PhLi	$PhCOCH_2CH_2CH_3$ (up to 90)	(24)
$n\text{-}C_{17}H_{34}CO_2Li$	$CH_2{=}CHLi$	$n\text{-}C_{17}H_{34}COCH{=}CH_2$ (70)	(25)
$PhCO_2Li$	$4\text{-}MeC_6H_4Li$	$4\text{-}MeC_6H_4COPh$ (up to 79)	$(26)^a$

[a] In certain cases prolonged reaction led to "scrambling" of aryl groups.

0.170 mol) is added dropwise during 30 min. The ice bath is then removed and stirring is continued at room temperature for 2 h. The reaction flask is fitted with a dip tube to enable the contents to be transferred by nitrogen pressure. The reaction mixture, a fine suspension, is agitated and transferred into a vigorously stirred mixture of concentrated hydrochloric acid (27 ml, 0.32 mol) and water (400 ml). The flask is rinsed with ether (100 ml), which is also added to the aqueous mixture. The resulting mixture is saturated with sodium chloride. The organic phase is separated, and the aqueous phase (which is alkaline) is extracted with ether (3 × 150 ml). The combined organic phases are dried ($MgSO_4$) and most of the solvent is distilled (40 cm Vigreux column). The pale yellow, liquid residue is distilled (10 cm Vigreux column), the fraction b.p. 57–60°/8 mmHg being collected as acetylcyclohexane (ca. 17.6 g, 93%), n_D^{26} ca. 1.4485.

References

1. L. M. Baigrie, H. R. Seiklay and T. Tidwell, *J. Am. Chem. Soc.* **107**, 5391 (1985); see also R. Häner, T. Laube and D. Seebach, *Ibid.* **107**, 5396 (1985).
2. J. Einhorn and J. L. Luche, *Tetrahedron Lett.* **27**, 501 (1986).
3. L. M. Baigrie, D. Lenoir, H. R. Seiklay and T. Tidwell, *J. Org. Chem.* **20**, 2105 (1985).
4. J. A. Beel and E. Vejvoda, *J. Am. Chem. Soc.* **76**, 905 (1954).
5. S. A. Lebedev, S. V. Ponomarev and I. F. Lutsenko, *J. Gen. Chem. USSR (Engl. Transl.)* **42**, 643 (1972).
6. J. J. Fitt and H. J. Gschwend, *J. Org. Chem.* **44**, 303 (1979).
7. D. A. Shirley and M. D. Cameron, *J. Am. Chem. Soc.* **74**, 664 (1952).
8. H. Gilman and P. R. van Ess, *J. Am. Chem. Soc.* **55**, 1258 (1933).
9. M. J. Jorgensen, *Org. React.* **18**, 1 (1970).
10. E. H. Vickery, C. E. Browne, D. L. Bymaster, T. K. Dobbs, L. L. Ansell and E. J. Eisenbraun, *Chem. Ind.* 954 (1977).
11. W. H. Glaze and C. M. Selman, *J. Org. Chem.* **33**, 1987 (1968).
12. See Section 3.1.2, Ref. (7).
13. See Section 6.2, Ref. (2).

14. L. Brandsma and H. D. Verkruijsse, *Synthesis of Acetylenes, Allenes and Cumulenes*, p. 32, Elsevier, Amsterdam, 1981.
15. A. C. Cope, M. Burg and S. W. Fenton, *J. Am. Chem. Soc.* **74**, 173 (1952).
16. R. A. Holton and H.-B. Kim, *Tetrahedron Lett.* **27**, 2191 (1986).
17. D. A. Shirley, J. R. Johnson and J. P. Hendrix, *J. Organomet. Chem.* **11**, 209 (1968).
18. N. Meyer and D. Seebach, *Chem. Ber.* **113**, 1304 (1980).
19. M. D. Rausch, G. A. Moser and C. F. Meade, *J. Organomet. Chem.* **51**, 1 (1973).
20. G. Köbrich and H. Trapp, *Chem. Ber.* **99**, 670 (1966).
21. T. M. Bare and H. O. House, *J. Org. Chem.* **33**, 943 (1968).
22. R. J. Boatman, B. J. Whitlock and H. W. Whitlock, *J. Am. Chem. Soc.* **99**, 4822 (1977).
23. M. Bogavac, L. Arsenijevic, S. Pavlov and V. Arseniyevic, *Tetrahedron Lett.* **25**, 1843 (1984).
24. R. Levine, M. J. Karten and W. M. Kadunce, *J. Org. Chem.* **40**, 1770 (1975).
25. J. C. Floyd, *Tetrahedron Lett.* 2877 (1974).
26. P. Hodge, G. M. Perry and P. Yates, *J. Chem. Soc. Perkin Trans. 1* 680 (1977).
27. T. M. Bare and H. O. House, *Org. Synth. Coll. Vol.* **5**, 775 (1973).

6.5. ADDITION TO CARBON MONOXIDE AND METAL CARBONYLS

6.5.1. Addition to Carbon Monoxide; Generation and Trapping of Acyllithium Compounds

It is well-established that the primary product from the reaction of carbon monoxide with an organolithium compound is an acyllithium compound:

$$\text{RLi} + \text{CO} \longrightarrow \left[\begin{array}{c} R \\ \diagdown \\ \diagup \\ Li \end{array} C=O \right]$$

However, until recently almost all attempts to utilize the primary adduct were unsuccessful, and although some secondary products were obtained in acceptable yields, such reactions were of very limited usefulness (for a review see Ref. (*1*)). One recent attempt that met with some success involved the reaction of phenyllithium with carbon monoxide *in the presence* of a primary alkyl halide RX at $-78°$ to give good yields of the tertiary alcohol $Ph_2C(OH)R$, derived from addition of phenyllithium to the ketone PhCOR (*2*). However, it is the work of Seyferth *et al.* that has demonstrated that under the right conditions many acyllithium compounds may be trapped in fair-to-good yields by a variety of electrophiles (*1*). The range of electrophilic reagents that have been employed, and the types of product obtained, are shown in Table 6.8. The success of these reactions depends on the organolithium compound reacting more rapidly with carbon monoxide than with

TABLE 6.8

Trapping of Acyllithium Intermediates (R^1COLi) by Electrophiles

Electrophilic reagent	Product[a]	Ref.
$R^2N=C=NR^2$	$R^1COC{\overset{NHR^2}{\underset{NR^2}{<}}}$	(3)
R^2CHO	$R^1COCH(OH)R^2$	(4)
R^2COR^3	$R^1COC(OH)R^2R^3$	(4, 5)
$R^2COOR^{3\,b}$	R^1COCOR^2	(6)
$R^2N=C=O$	$R^1COCONHR^2$	(7)
$Fe(CO)_5{}^c$	$R^1CO\underset{OLi}{\overset{\mid}{C}}:Fe(CO)_4$	(8)
$R^2N=C=S$	$R^1COCSNHR^2$	(7)
CS_2	$R^1COS^{\ominus\,d}$	(9)
R^2SSR^2	R^1COSR^2	(10)
S_n	R^1COSMe^e	(10)
R^2_3SiCl	R^1COSiR^3	(11)

[a] After hydrolysis where appropriate. [b] Also with lactones (5). [c] See Section 6.5.2. [d] By loss of CS from R^1COCS_2Li; trapped by MeI, Me_3SiCl or Me_3SnBr. [e] After reaction with MeI.

the co-electrophile at the very low temperatures used—as low as $-135°$. Alkyllithium compounds have so far proved most successful in these reactions, especially where there is some steric hindrance. These types of reaction are clearly capable of further development; in particular the use of different types of organolithium compound requires further study (12).

Analogous reactions of cuprates, giving conjugate addition of the acyl-metal intermediate to enones, have also been reported (13):

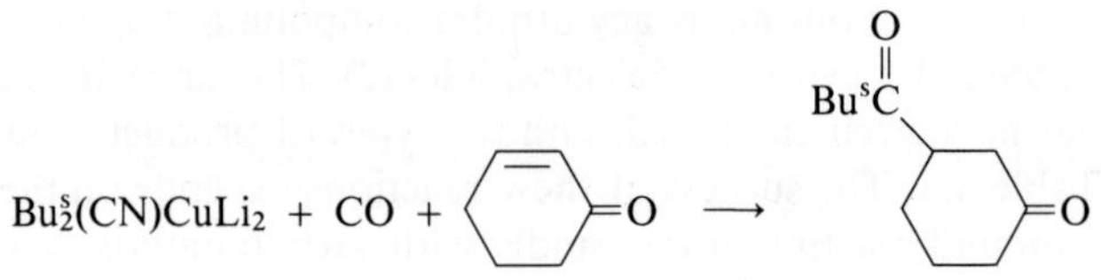

$Bu^s_2(CN)CuLi_2 + CO + $ [cyclohexenone] $\longrightarrow$ [3-(Bu^sCO)-cyclohexanone]

6.5. ADDITION TO CARBON MONOXIDE AND METAL CARBONYLS

Generation and trapping of an acyllithium compound;
3-hydroxy-2,2,3-trimethyloctan-4-one (14)

$$CO + Me_3CCOMe \xrightarrow[\text{(ii) aq. NH}_4\text{Cl}]{\text{(i) BuLi, } -110°} BuCOC(OH)CMe_3$$
$$\qquad\qquad\qquad\qquad\qquad\quad |$$
$$\qquad\qquad\qquad\qquad\qquad\ Me$$

A 2 l three-necked flask is equipped with a stirrer, a low-temperature thermometer and a gas dispersion tube. In the flask are placed THF (400 ml), ether (100 ml), pentane (100 ml) and pinacolone (3,3-dimethylbutan-2-one; distilled from potassium carbonate) (7.92 g, 79.1 mmol). The mixture is cooled to $-110°$[a] and carbon monoxide is bubbled in for 30 min. n-Butyllithium (ca. 2.5 M in pentane, 78.4 mmol) is added at 0.67 ml min^{-1} by means of a syringe pump.[b] The stream of carbon monoxide is maintained while the orange reaction mixture is stirred at $-110°$ for 2 h. The mixture is allowed to warm to room temperature over 1.5 h, during which time it becomes yellow in colour. Saturated aqueous ammonium chloride is added. The colourless aqueous layer is separated from the pale-yellow organic layer and washed with pentane (2 × 100 ml). The organic layers are combined and dried (MgSO$_4$), and the solvents are removed by distillation through a short Vigreux column. The residue is distilled under reduced pressure, the fraction b.p. 120–122°/30 mmHg being collected as 3-hydroxy-2,2,3-trimethyloctan-4-one (10.8 g, 73%), n_D^{20} 1.4442.

[a] It is important to measure the true internal temperature of the reaction mixture.
[b] If a syringe pump is not available then very slow manual addition at as constant a rate as possible may be employed.

References

1. D. Seyferth, R. M. Weinstein, W.-L. Wang, R. C. Hui and C. M. Archer, *Israel J. Chem* **24**, 167 (1984).
2. N. S. Nudelman and A. A. Vitale, *J. Org. Chem.* **46**, 4625 (1981); A. A. Vitale, F. Doctorovich and N. S. Nudelman, *J. Organomet. Chem.* **332**, 9 (1987).
3. D. Seyferth and R. C. Hui, *J. Org. Chem.* **50**, 1985 (1985).
4. D. Seyferth, R. M. Weinstein, W.-L. Wang and R. C. Hui, *Tetrahedron Lett.* **24**, 4907 (1983).
5. R. M. Weinstein, W.-L. Wang and D. Seyferth, *J. Org. Chem.* **48**, 3367 (1983).
6. D. Seyferth, R. M. Weinstein and W.-L. Wang, *J. Org. Chem.* **48**, 1144 (1983).
7. D. Seyferth and R. C. Hui, *Tetrahedron Lett.* **25**, 5251 (1984).
8. K. H. Dötz, U. Wenicker, G. Müller, H. G. Alt and D. Seyferth, *Organometallics* **5**, 2570 (1986).
9. D. Seyferth and R. C. Hui, *Tetrahedron Lett.* **25**, 2623 (1984).
10. D. Seyferth and R. C. Hui, *Organometallics* **3**, 327 (1984).
11. D. Seyferth and R. M. Weinstein, *J. Am. Chem. Soc.* **104**, 5534 (1982).

12. D. Seyferth, W.-L. Wang and R. C. Hui, *Tetrahedron Lett.* **25,** 1651 (1984).
13. D. Seyferth and R. C. Hui, *J. Am. Chem. Soc.* **107,** 455 (1985).
14. R. Hui and D. Seyferth, Personal communication.

6.5.2. Addition to Metal Carbonyls

The addition of organolithium compounds to transition metal carbonyls generally follows the pattern shown here:

$$\text{RLi} + \text{M(CO)}_n \longrightarrow \underset{\underset{\text{OLi}}{|}}{\text{R}-\text{C}:\text{M(CO)}_{n-1}}$$

This type of reaction, and the chemistry of the resulting "carbene complexes", have been extensively studied (*1, 2*), but their application in organic synthesis has developed slowly, despite the early realization that they could act as acyl-anion equivalents (*1*):*

$$\underset{\underset{\text{O}^{\ominus}\text{Li}^{\oplus}}{|}}{\overset{\overset{\text{E}^{\oplus}}{|}}{\text{R}-\text{C}:\text{M(CO)}_{n-1}}} \longrightarrow \text{R}-\text{C}\overset{\nearrow\text{E}}{\underset{\diagdown}{\diagdown}}\text{O}$$

Many applications involve, for example, methoxycarbene complexes obtained by *O*-methylation of the initial adducts, and thus fall outside the scope of a book on organolithium methods. The preparation of a methoxycarbene complex described below is typical. This is adapted from one of a group of three papers which briefly review earlier work as well as reporting new results

---

*An ingenious extension of this principle is shown below; addition to coordinated carbon monoxide is followed by intramolecular conjugate addition to coordinated enone (*4*):

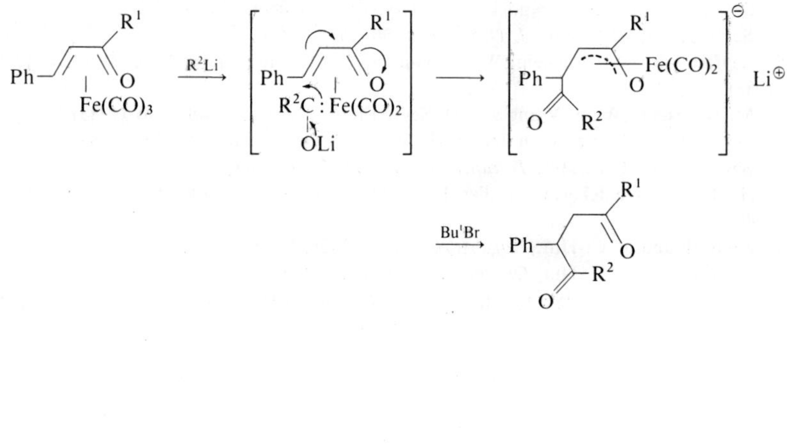

6.5. ADDITION TO CARBON MONOXIDE AND METAL CARBONYLS

on the synthesis of anthracyclinones (4–6). The addition of methyllithium to chromium hexacarbonyl (7) and of methyl- and phenyllithium to a carbonyl group in the complex [Ru$_2$(CO)$_4$(η-C$_5$H$_5$)$_2$] have been described in detail (8).

Addition to chromium hexacarbonyl; pentacarbonyl[methyoxy(1,4-dimethoxy-2-naphthyl)carbene] chromium(0) (4)

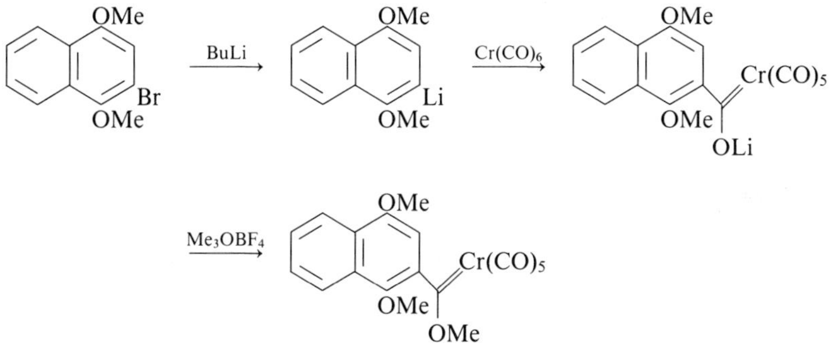

2-Bromo-1,4-dimethoxynaphthalene (4.01 g, 15 mmol) is dissolved in diethyl ether (40 ml) under nitrogen. The solution is stirred at room temperature as n-butyllithium (*ca.* 1.6 M in hexane, 15 mmol) is added during *ca.* 45 min, during which time the colour changes from yellow to orange. Stirring is continued for 2 h.

Chromium hexacarbonyl (3.30 g, 15 mmol) is suspended in ether (75 ml) under nitrogen, and the suspension is stirred as the naphthyllithium solution is added dropwise during 30 min. The suspension first turns lemon-yellow, then reddish-brown. Stirring is continued for 2 h, and the solvent is evaporated under reduced pressure to leave a yellowish powder. This powder is dissolved in water (100 ml). Pentane (100 ml) and trimethyloxonium tetrafluoroborate (3.34 g, 22.5 mmol) are immediately added and the mixture is stirred vigorously. The pentane layer is separated and the aqueous layer is extracted with pentane (5 × 100 ml). The combined extracts (which are deep-red) are dried (Na$_2$SO$_4$) and evaporated. Chromatography of the residue at −25° (silica, dichloromethane–pentane) gives the methoxycarbene complex as red crystals (4.7 g, 74%), m.p. 93°, which are stable in air for some time.

References

1. M. Ryang, *Organomet. Chem. Rev. (A)* **5**, 67 (1970).
2. E. O. Fischer, *Adv. Organomet. Chem.* **14**, 1 (1976).
3. S. E. Thomas, *J. Chem. Soc. Chem. Commun.* 226 (1987).

4. K. H. Dötz and M. Popall, *Tetrahedron* **41**, 5797 (1985).
5. M. F. Semmelhack, J. J. Bozell, L. Keller, T. Sato, E. J. Spiess, W. Wulff and A. Zask, *Tetrahedron* **41**, 5803 (1985).
6. W. D. Wulff, P.-C. Tang, K.-S. Chan, J. S. McCallum, D. C. Yang and S. R. Gilbertson, *Tetrahedron* **41**, 5813 (1985).
7. L. S. Hegedus, M. A. McGuire and L. M. Schultze, *Org. Synth.* **65**, 140 (1987).
8. N. M. Doherty and S. A. R. Knox, *Organomet. Synth.* **3**, 222 and 229 (1986).

—7—
Addition of Organolithium Compounds to Thiocarbonyl Groups

Although some reactions of organolithium compounds with thiocarbonyl compounds are analogous to those with carbonyl compounds, formally involving addition by attack of the "carbanion" on the thiocarbonyl carbon, an alternative mode of addition is often observed, in which the "carbanion" bonds to sulphur. The former mode of addition is known as carbophilic, and the latter, thiophilic.

$$R^1Li + R^2R^3C=S \longrightarrow \underset{\underset{Li}{|}}{R^2R^3C-SR}$$

It is not possible to predict with certainty which mode of addition will be followed in a particular case, though some broad generalizations can be made. The mechanism of thiophilic addition is also unclear, though some electron-transfer/radical character seems likely. Only representative types of both modes of reaction are included here; for more detailed discussion see General Ref. D(ii), pp. 40 and 68.

7.1. CARBOPHILIC ADDITION

Additions of organolithium compounds to thiocarbonyl groups, to form carbon–carbon bonds, are not on the whole of great practical value. The exception is addition to cumulated thiocarbonyl groups, and particularly to carbon disulphide, which is a useful method for preparing dithiocarboxylates and compounds that can be made from them *in situ*:

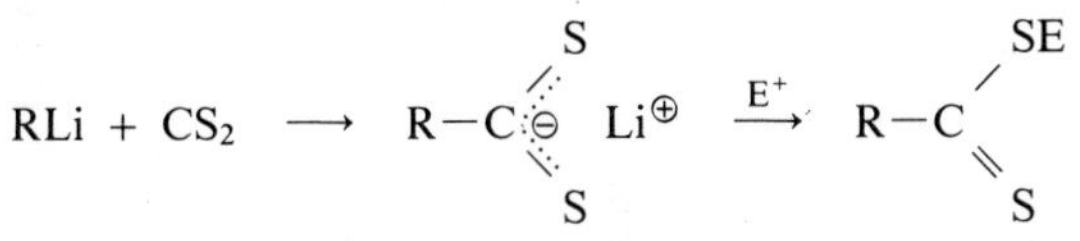

$$RLi + CS_2 \longrightarrow R-C\underset{S}{\overset{S}{\diagup}}\ Li^\oplus \xrightarrow{E^+} R-C\underset{S}{\overset{SE}{\diagup}}$$

101

TABLE 7.1

Reactions of Organolithium Compounds with Carbon Disulphide[a] and with Isothiocyanates

Thiocarbonyl reagent	Organolithium compound	Further reagent	Product (yield %)	Ref.
CS_2	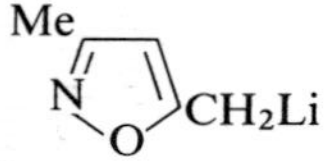 (3-methylisoxazol-5-yl)CH_2Li	MeI	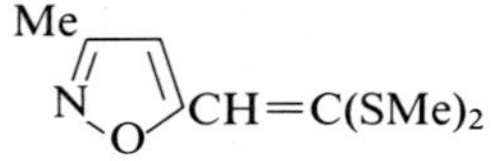 3-methyl-5-(isoxazolyl)$CH=C(SMe)_2$ (75)	(1)
CS_2	$LiCH_2NC$	MeI	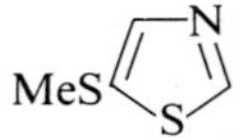 MeS-thiazolyl (b) (34)	(2)
CS_2	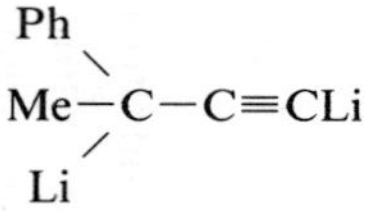 $PhMeC(Li)-C\equiv CLi$	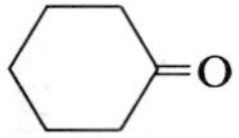 cyclohexanone	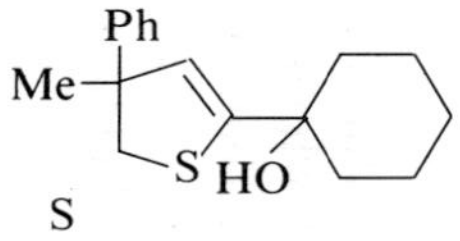 4-Ph-4-Me-dihydrothiophene with cyclohexyl-HO, S (81)	(3)[b]
CS_2	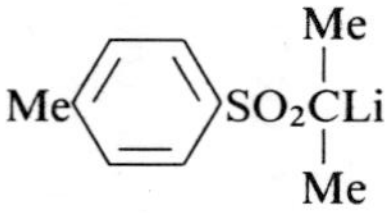 2-($MeSO_2CMe_2Li$)-phenyl	$BrCH_2COOEt$	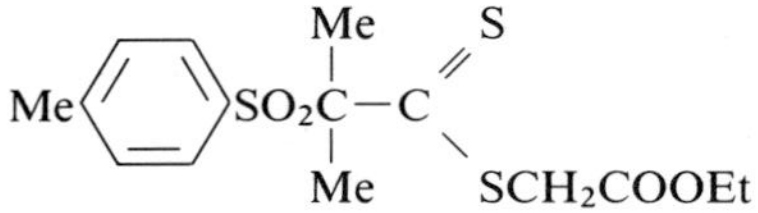 $Me-C_6H_4-SO_2-CMe_2-C(=S)-SCH_2COOEt$ (41)	(4)
MeNCS	Bu^tLi	H_3O^+	$MeNHCSBu^t$ (89)	(5)[c]
$MeOCH_2NCS$	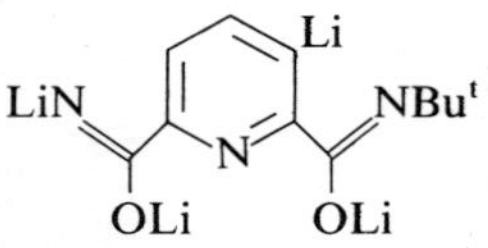 2,6-bis(LiN=C(OLi)-)pyridine with Bu^t	H_3O^+	 2-($CSNHCH_2OMe$)-6-($CONHBu^t$)-pyridine (88)	(6)
PhNCS	Bu^nLi	H_3O^+	$PhNHCSBu^n$ (88)	(5)

[a] Copper(I) catalysis is reported to improve yields of reactions of aryllithium compounds with carbon disulphide (6a). [b] By cyclization of

7.1. CARBOPHILIC ADDITION

Addition to isothiocyanates may also be used for synthesizing thioamides:

$$R^1Li + R^2N=C=S \longrightarrow R^2N=C\begin{smallmatrix}R^1\\ \\SLi\end{smallmatrix} \xrightarrow{H_3O^+} R^2NHCSR^2$$

Examples of both types of reaction are listed in Table 7.1. The detailed example of a reaction with carbon disulphide involves the lithium enolate of a carboxylate.

Reaction with carbon disulphide; methyl 2-carboxydithiohexadecanoate (7)

$$C_{14}H_{29}CH_2CO_2H \xrightarrow{2\ LDA} C_{14}H_{29}CH=C(OLi)_2 \xrightarrow[\text{(ii) MeI}]{\text{(i) CS}_2} C_{14}H_{29}CH\begin{smallmatrix}CO_2H\\ \\CSSMe\end{smallmatrix}$$

Diisopropylamine (2.74 ml, 21 mmol) is added to THF (75 ml) contained in a flask flushed with nitrogen and cooled to $-50°$. The temperature is maintained at $-50°$ as n-butyllithium (*ca.* 1.6 M in hexane, 21 mmol) is added, followed by a solution of hexadecanoic acid (2.56 g, 10 mmol) and hexametapol (1.8 ml, 10 mmol) in THF. The solution is heated to $35°$ for 30 min, then cooled to $-30°$ as carbon disulphide (0.665 ml, 11 mmol) is added. The temperature is maintained at $-30°$ for 10 min, then lowered to $-50°$. Iodomethane (0.619 ml, 10 mmol) is added, and the temperature is maintained at $-50°$ for 30 min. While the temperature is still at $-50°$, dilute hydrochloric acid is added, causing a rapid rise in temperature. The aqueous layer is separated and extracted three times with light petroleum. The combined organic layers are washed with dilute hydrochloric acid and then with water, dried (Na_2SO_4) and evaporated at room temperature to leave a yellow solid (3.30 g). Recrystallization from chloroform (20 ml) gives methyl 2-carboxydithiohexadecanoate (2.54 g, 73%), m.p. 67–72° dec.

In another experiment, following the addition of carbon disulphide the solution was heated to $50°$ for 2 h, resulting in decarboxylation. Addition of iodomethane followed by further heating gave a product consisting mainly of 1,1-di(methylthio)hexadec-1-ene and methyl dithiohexadecanoate.

In contrast with vinylogous amides, where conjugate addition of organolithium compounds is followed by elimination of lithium dialkylamide (see Section 6.3, p. 86), vinylogous thioamides give 1,4-adducts, which may be trapped by methylation as follows (8):

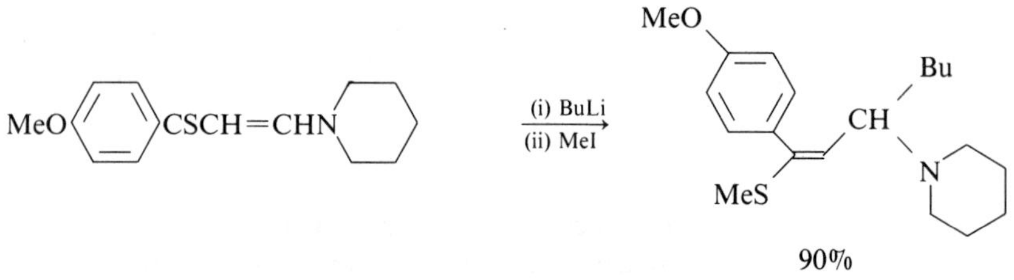

MeO—⟨C₆H₄⟩—CSCH=CHN⟨piperidine⟩ $\xrightarrow[\text{(ii) MeI}]{\text{(i) BuLi}}$ MeO—⟨C₆H₄⟩—C(MeS)=CH—CH(Bu)—N⟨piperidine⟩

90%

Vinylogous dithiocarbamates react similarly.

References

1. R. G. Micetich, *Can. J. Chem.* **48**, 2006 (1970).
2. U. Schöllkopf, P.-H. Porsch and E. Blume, *Justus Liebigs Ann. Chem.* 7122 (1976).
3. A Commercon and G. Ponsinet, *Tetrahedron Lett.* **26**, 5131 (1985); see also J. Meijer, K. Ruitenberg, H. Westmijze and P. Vermeer, *Synthesis* 551 (1981).
4. M. van der Leij, H. J. M. Strijtveen and B. Zwanenburg, *Rec. Trav. Chim. Pays-Bas* **99**, 45 (1980).
5. G. Entenmann, *Chem.-Ztg* **101**, 508 (1977).
6. T. R. Kelly, A. Echavarren, N. S. Chandrakumar and Y. Köksal, *Tetrahedron Lett.* **25**, 2127 (1984).
6a. H. D. Verkruijsse and L. Brandsma, *J. Organomet. Chem.* **332**, 95 (1987).
7. D. A. Konen, P. E. Pfeffer and L. S. Silbert, *Tetrahedron* **32**, 2705 (1976).
8. J.-P. Guémas, M. Lees, A. Reliquet, F. Reliquet and H. Quiniou, *Phosphorus Sulfur* **8**, 351 (1980); J.-P. Guémas, M. Lees, A. Reliquet and J. Villieras, *Ibid.* **12**, 325 (1982).

7.2. THIOPHILIC ADDITION

Both the mechanisms and the applications of thiophilic addition to thiocarbonyl compounds have been more thoroughly investigated for Grignard reagents than for organolithium compounds (see General Ref. D(ii)). Nevertheless, it is established that high yields of thiophilic addition products are obtainable with the latter, as in the reaction of aryllithium compounds or n-butyllithium with thiobenzophenone (*1, 2*):*

$$Ph_2C=S \xrightarrow[\text{(ii) H}_2\text{O}]{\text{(i) RLi}} Ph_2CHSR$$

With other thioketones, however, some carbophilic addition may also occur,

*Thiophilic reactions with phenyl dithiobenzoate and phenyl trithiocarbonate are also described (*1*).

7.2. THIOPHILIC ADDITION

and in many cases the main reaction is reduction to the corresponding thiol (by β-hydrogen transfer) (2-4). This reduction reaction is made use of in an overall transformation of a ketone into the corresponding thiol, an example of which is described in *Organic Syntheses* (5):

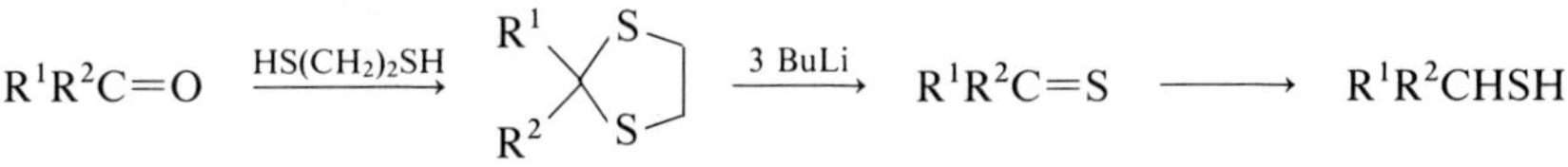

$$R^1R^2C=O \xrightarrow{HS(CH_2)_2SH} \underset{R^2}{\overset{R^1}{>}}\!\!\!\!\underset{S}{\overset{S}{<}} \xrightarrow{3\ BuLi} R^1R^2C=S \longrightarrow R^1R^2CHSH$$

Thiophilic additions of organolithium compounds to thioketenes give lithiated vinyl sulphides, which can be trapped by electrophiles at low temperatures; at higher temperatures carbenoid reactions occur (6):

$$R^1R^2C=C=S \xrightarrow{R^3Li} R^1R^2C=C\underset{Li}{\overset{SR^3}{<}} \xrightarrow{E^+} R^1R^2C=C\underset{E}{\overset{SR^3}{<}}$$

References

1. P. Beak and J. W. Worley, *J. Am. Chem. Soc.* **94**, 597 (1972).
2. A. Ohno, K. Nakamura, M. Uohama and S. Oka, *Chem. Lett.* 983 (1975).
3. A. Ohno, K. Nakamura, M. Uohama, S. Oka, T. Yamabe and S. Nagata, *Bull. Chem. Soc. Jpn* **48**, 3718 (1975).
4. V. Rautenstrauch, *Helv. Chim. Acta* **57**, 496 (1974).
5. S. R. Wilson and G. M. Georgiadis, *Org. Synth.* **61**, 74 (1983).
6. E. Schaumann and W. Walter, *Chem. Ber.* **107**, 3562 (1974).

—8—

Substitution at Carbon by Organolithium Compounds

The reactions described here are formally nucleophilic substitutions at carbon by "carbanions", though their mechanisms are in many cases far from straightforward. Organic halides are most commonly used as substrates, but in view of the complications often encountered, alternative leaving groups such as sulphonate have been tried, and have been found useful in a few cases. Alkoxides are not usually sufficiently good leaving groups except in the special case of ring opening of oxiranes and oxetanes. The useful reactions of acetals and orthoesters with Grignard reagents are rarely satisfactory with organolithium compounds (see General Ref. D(ii)).

8.1. DISPLACEMENT OF HALIDE

The following desirable reaction is susceptible to side-reactions involving elimination and metal–halogen exchange, and its mechanism and stereochemistry are often complicated (see General Ref. A and Ref. (*1*)):

$$R^1Li + R^2X \longrightarrow R^1R^2 + LiX$$

Nevertheless, good results are usually obtained without difficulty with organolithium compounds in which the "carbanion" is delocalized and primary alkyl halides. Bromides and iodides are generally preferred, but iodides are more susceptible to metal–halogen exchange, though methyl iodide usually works well. Benzylic and allylic halides are often satisfactory, though the former (particularly the chlorides) are susceptible to carbenoid reactions and with the latter rearranged products may be formed by S_N2'-type reactions.

In cases of difficulty, variations in the solvent and other conditions sometimes lead to success; the presence of additives such as hexametapol in the reaction medium is often beneficial, as in the synthesis of 3-butylpyridine described below.

Vinylic and aromatic halides are only directly substituted by organolithium compounds in special cases (see General Refs. A and D(ii)), though metal–halogen exchange followed by coupling or reactions via arynes (see Section 16.1) may give the same overall result:

$$\text{ArX} + \text{RLi} \rightleftharpoons \text{ArLi} + \text{RX} \longrightarrow \text{ArR} + \text{LiX}$$

When alkylation of these types of halide* is desired, the use of organocopper intermediates (which may be prepared *in situ* from the organolithium compound) is the method of choice (*2, 3*). Palladium-induced cross-coupling with vinylic and aromatic halides has been applied to Grignard reagents more than to organolithium compounds, but examples involving the latter are known and a general procedure has been described (*4*).

Many examples of reactions of organolithium compounds with organic halides are tabulated in General Ref. A, and further well-described examples are listed in Table 8.1. Two representative alkylations of primary alkyl halides are fully described.

Synthesis of 3-butylpyridine from 3-(lithiomethyl)pyridine and 1-bromopropane (14)

<chemical structures: 3-methylpyridine → (LDA) → 3-(lithiomethyl)pyridine → (PrBr) → 3-butylpyridine>

Diisopropylamine (2.53 g, 25 mmol) and THF (10 ml) are placed in a 300 ml three-necked flask equipped with a pressure-equalizing dropping funnel, a rubber septum and a magnetic stirrer, and furnished with an inert atmosphere. The mixture is cooled to 0° and stirred as n-butyllithium (*ca.* 1.6 M in hexane, 25 mmol) is added by syringe. The resulting pale-yellow solution is stirred at 0° for 15 min. Hexametapol (4.5 g, 25 mmol) in THF (10 ml) is added and the resulting bright-yellow solution is stirred at 0° for 15 min. A solution of 3-methylpyridine (2.3 g, 25 mmol) in THF (10 ml) is added during 5 min. Stirring is continued at 0° for 30 min, and a solution of 1-bromopropane (3.10 g, 25 mmol) in THF (15 ml) is added. The resulting solution is stirred at 25° for 1 h, and then poured into 10% hydrochloric acid. The resulting two layers are separated. The aqueous layer is made basic by the addition of solid potassium hydroxide and extracted with ether (3 × 20 ml). The combined organic layers are washed with water, dried

---

* Organocopper intermediates may also be used for displacement of halogen in the presence of a carbonyl group (*2, 3*).

TABLE 8.1

Alkylation of Organic Halides by Organolithium Compounds

Halide	Organolithium compound	Product (yield %)	Ref.
MeI	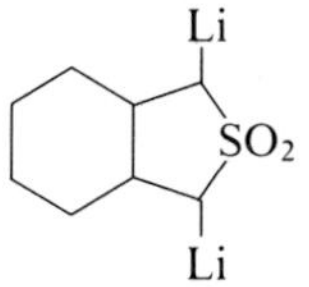 2-lithio-octahydrobenzo[c]thiophene 1,1-dioxide (bicyclic, Li on C-α to SO_2)	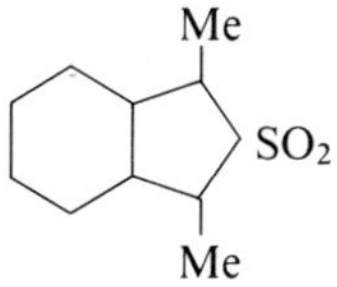 2-methyl analogue, SO_2 (87–89.5)	(5)
MeI	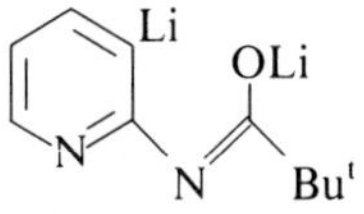 3-Li-2-(N-Li, N'-Bu^t-amidino)pyridine	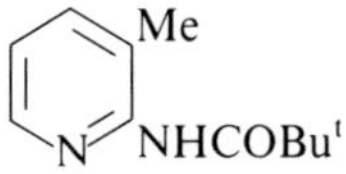 3-Me-2-(NHCOBu^t)pyridine (up to 89)	(6)
EtBr	Pr^{n}CHC≡CLi, with Li on CH	Pr^{n}CH(Et)C≡CH (64–65)	(7)
n-C$_6$H$_{13}$Br	$H_2C=C=C$(OMe)(Li)	$H_2C=C=C$(OMe)(C_6H_{13}) (85)	(8)
n-C$_{14}$H$_{29}$Br	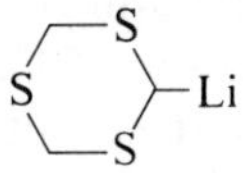 2-Li-1,3-dithiane	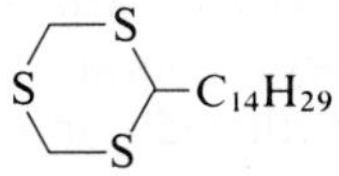 2-$C_{14}H_{29}$-1,3-dithiane (almost quant.)	(9)
Br(CH$_2$)$_4$Br	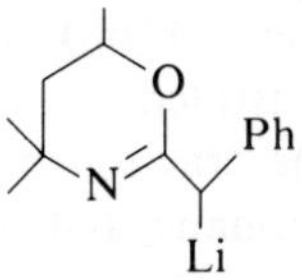 4,4-dimethyl-2-(α-Li-benzyl)-5,6-dihydro-1,3-oxazine	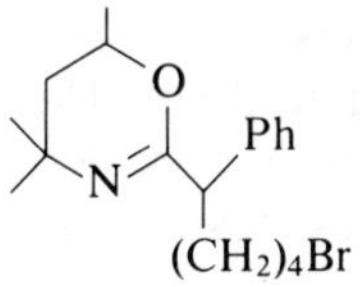 4,4-dimethyl-2-[CHPh(CH$_2$)$_4$Br]-5,6-dihydro-1,3-oxazine (almost quant.)	(10)
EtOCH$_2$Cl	$H_2C=C=C$(OMe)(Li)	$H_2C=C=C$(OMe)(CH$_2$OEt) (78)	(11)
Me$_2$CHI	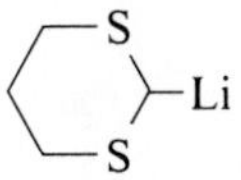 2-Li-1,3-dithiane	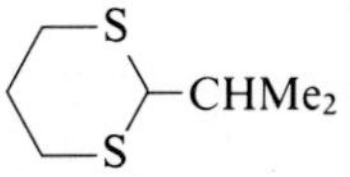 2-CHMe$_2$-1,3-dithiane (82)	(12)
Me$_2$C=CHCH$_2$Br	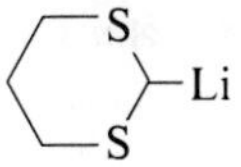 2-Li-1,3-dithiane	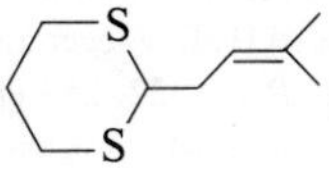 2-(CH$_2$CH=CMe$_2$)-1,3-dithiane (71)	(12)
PhCH$_2$Cl	PhCH(SMe)(Li)	PhCH(SMe)(CH$_2$Ph) (66)	(13)

(calcium chloride) and concentrated. Distillation of the residue gives 3-butylpyridine (2.60 g, 77%), b.p. 74–76°/7 mmHg.

Syntheses via lithiated 1,3-dithianes; 5,9-dithiaspiro[3.5]nonane (15)

$$\left[\begin{array}{c}\overset{S}{\underset{S}{\diagup}}\!\!\!\diagdown\!\text{Li}\end{array}\right] \xrightarrow{\text{Br(CH}_2)_3\text{Cl}} \left[\begin{array}{c}\overset{S}{\underset{S}{\diagup}}\!\!\!\diagdown\!\text{(CH}_2)_3\text{Cl}\end{array}\right] \xrightarrow{\text{BuLi}} \left[\begin{array}{c}\overset{S}{\underset{S}{\diagup}}\!\!\!\diagdown\!\!\diagdown\!\!\!\underset{\text{Li Cl}}{\diagdown}\end{array}\right]$$

$$\longrightarrow \overset{S}{\underset{S}{\diagup}}\!\!\!\diagdown\!\!\diagdown\!\!\!\diagdown$$

A solution of 2-lithio-1,3-dithiane is prepared as described on p. 36 and cooled to $-75°$ (bath temperature). 1-Bromo-3-chloropropane (65.5 g, 44.5 ml, 0.417 mol) is added by syringe during 10 min. The bath temperature is slowly raised to $-30°$ during 2 h, and then to room temperature during a further 2 h. The cooling bath is again cooled to $-75°$ and n-butyllithium (in hexane, 0.44 mol) is added by syringe during 10 min, and the reaction mixture is then allowed to warm to room temperature overnight. The solvent is evaporated (rotary evaporator, bath temperature 50°, *ca.* 20 mmHg), and water (300 ml) and ether (500 ml) are added to the residue. The organic layer is separated, and the aqueous layer is extracted with ether (500 ml). The combined organic layers are washed with water (200 ml) and dried (anhydrous potassium carbonate, 10 g). The ether is distilled and the residue (*ca.* 75 g) is distilled through a packed column, the fraction b.p. 65–75°/1 mmHg being collected as 5,9-dithiaspiro[3.5]nonane (44–57 g, 65–84%), n_D^{20} 1.5700.

The *Organic Syntheses* description includes the hydrolysis of the product to cyclobutanone (*15*).

References

1. I. P. Beletskaya, G. A. Artamkina and O. A. Reutov, *Russ. Chem. Rev. (Engl. Transl.)* **45**, 330 (1976); see also H. E. Zieger and D. Mathison, *J. Am. Chem. Soc.* **101**, 2207 (1979).
2. G. H. Posner, *Org. React.* **22**, 253 (1975).
3. See Section 6.1.2, Ref. (*3b*).
4. R. F. Heck, *Palladium Reagents in Organic Syntheses*, Academic Press, London, 1985, Chap. 6; see also Section 3.2, Ref. (*52*).
5. J. M. Photis and L. A. Paquette, *Org. Synth.* **57**, 53 (1977).
6. See Section 3.2, Ref. (*7*).
7. See Section 3.2, Ref. (*41*).

8. See Section 6.4, Ref. (*14*), p. 38.
9. D. Seebach and A. K. Beck, *Org. Synth,* **51**, 39 (1971).
10. I. R. Politzer and A. I. Meyers, *Org. Synth.* **51**, 24 (1971).
11. See Section 6.4, Ref. (*14*), p. 40.
12. See Section 6.2, Ref. (*2*).
13. See Section 3.2, Ref. (*46*).
14. E. M. Kaiser and J. D. Pettie, *Synthesis* 705 (1975).
15. See Section 3.2, Ref. (*29*).

8.2. DISPLACEMENT OF SULPHATE AND SULPHONATE

On the whole, the use of leaving groups other than halogen offers little or no advantage, but there are a few exceptions to this generalization.

(a) Reaction with dimethyl sulphate is a convenient alternative to reaction with iodomethane. Some examples are listed in Table 8.2, and others are tabulated in General Ref. A. Diethyl sulphate has also been used (*1*).

TABLE 8.2

Reactions of Organolithium Compounds with Dimethyl Sulphate

Organolithium compound	Product (yield %)	Ref.
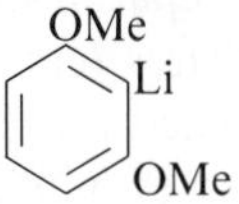 2,6-dimethoxyphenyllithium	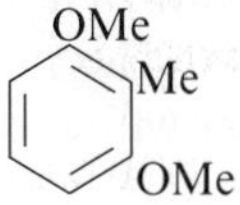 1,3-dimethoxy-2-methylbenzene (74–76)	(2)
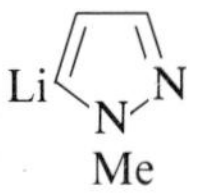 1-methyl-2-lithioimidazole	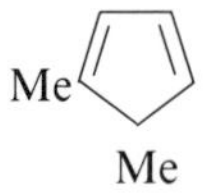 1-methyl-2-methylimidazole (75)	(3)
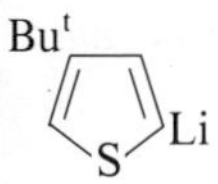 5-tert-butyl-2-lithiothiophene	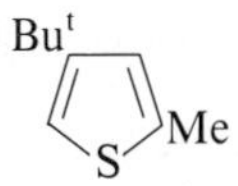 5-tert-butyl-2-methylthiophene (72)	(4)^a
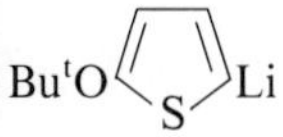 5-tert-butoxy-2-lithiothiophene	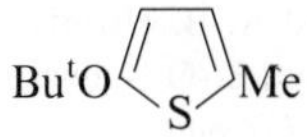 5-tert-butoxy-2-methylthiophene (87)	(5)
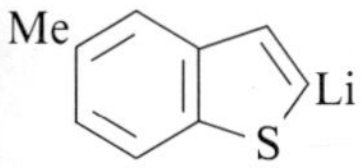 4-methyl-2-lithiobenzothiophene	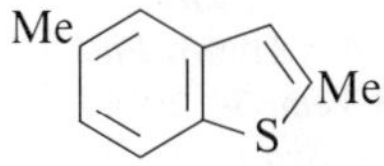 4-methyl-2-methylbenzothiophene (85)	(6)

^{a} Several other examples described; methylation used to determine orientation of metallation and metal–halogen exchange reactions.

(b) Reaction with sulphonate esters is also often satisfactory, and in one case has been reported to proceed with 93% inversion of configuration (7).* Most work has been carried out with tosylates, but the originators of the experiment described below report that the use of benzenesulphonates avoids side-reactions involving lithiation of the ring methyl group in tosylates. An important limitation is that many reactions of organolithium compounds with arylsulphonates proceed by thiophilic attack (displacement of alkoxide) rather than by substitution at carbon (9).

Alkylation of an arenesulphonate; 2-ethyl-1,3-dithiane (10)

This procedure has been used for several analogous reactions.

$$\underset{S}{\overset{S}{\diagdown}}\!\!\!\diagup\!\!\text{Li} + \text{EtOSO}_2\text{Ph} \longrightarrow \underset{S}{\overset{S}{\diagdown}}\!\!\!\diagup\!\!\text{Et}$$

The experimental procedure for preparing a solution of 2-lithio-1,3-dithiane is as described in Section 3.2, p. 36, but using 1,3-dithiane (1.20 g, 10 mmol) in THF (20 ml). To this solution is added with stirring a solution of ethyl benzenesulphonate (18.6 g, 10 mmol) in THF (8 ml). The cooling bath is removed and stirring is continued for 24 h. A colourless precipitate separates from the slightly yellow solution. The mixture is poured into water (ca. 150 ml) and extracted several times with pentane. The extract is washed with aqueous potassium hydroxide, dried (K_2CO_3) and evaporated, to leave crude 2-ethyl-1,3-dithiane (85%), b.p. 83°/14 mmHg.

References

1. E. N. Karaulova, D. Sh. Meilanova and G. D. Gal'pern, *J. Gen. Chem. USSR (Engl. Transl.)* **30**, 3262 (1960).
2. J. P. Lambooy, *J. Am. Chem. Soc.* **78**, 771 (1976).
3. R. Hüttel and M. E. Schön, *Justus Liebigs Ann. Chem.* **625**, 55 (1959).
4. S. Gronowitz, B. Cederhund and A.-B. Hörnfeldt, *Chem. Scripta* **5**, 217 (1974).
5. H. J. Jakobsen, E. H. Larsen and S.-O. Lawesson, *Tetrahedron* **19**, 1867 (1963).
6. R. P. Dickinson and B. Iddon, *J. Chem. Soc. (C)* 182 (1971).
7. W. D. Korte, L. Kinner and W. C. Kaska, *Tetrahedron Lett.* 603 (1970).
8. T. Minami, S. Tokumasu and I. Hirao, *Bull. Chem. Soc. Jpn* **58**, 2139 (1985).
9. W. H. Baarschers, *Can. J. Chem.* **54**, 3056 (1976).
10. D. Seebach and E.-M. Wilka, *Synthesis* 476 (1976).

* Some variation in regioselectivity in reactions of allyllithium compounds with allylic halides, tosylates and phosphates has been reported (8).

8.3. NUCLEOPHILIC RING OPENING OF EPOXIDES AND OTHER CYCLIC ETHERS

The reaction of organolithium compounds with oxirane (ethylene oxide) itself is a very useful and general method for two-carbon homologation:

$$RLi + \underset{\triangle}{\overset{O}{}} \longrightarrow RCH_2CH_2OLi \xrightarrow{H_3O^+} RCH_2CH_2OH$$

An example is described below, and others are listed in Table 8.3; further examples are tabulated in General Ref. A.

The corresponding reaction with substituted oxiranes is more susceptible to side-reactions (see General Ref. D(ii)), notably those consequent on deprotonation, and while monosubstituted oxiranes often give reasonable yields, more highly substituted oxiranes usually present problems except with "delocalized" organolithium compounds. Two approaches have had some success in overcoming these problems. The use of cuprates (1), particularly cyanocuprates (2), is covered in the reviews cited. Recently it has been demonstrated that activation of the oxirane ring by complexation with boron trifluoride can lead to high yields of desired ring-opened products (3, 4); although this procedure is so far relatively untried, it is potentially so useful that an example is described below.

The regiochemistry of attack by organolithium compounds on substituted oxiranes is not always predictable. Usually the more electron-deficient carbon is attacked (as in the examples involving methyloxirane shown in Table 8.3), but steric factors are also important. The stereochemistry of the ring opening is normally *trans*.

Reaction with oxirane; (E)-4-ethyloct-3-en-1-ol (14)

$$\underset{n-C_4H_9}{\overset{Et}{}}\!\!\!\!\!\!\!\!\diagdown\!\!\!\!\!\!\!\!\!\diagup\!\!\!\!\!\!\!\!\overset{I}{} C=CH \quad \xrightarrow[\underset{\text{(iii) NH}_4\text{Cl aq.}}{\overset{\text{(i) RLi}}{\text{(ii)} \triangle}}]{} \quad \underset{n-C_4H_9}{\overset{Et}{}}\!\!\!\!\!\!\!\!\diagdown\!\!\!\!\!\!\!\!\!\diagup\!\!\!\!\!\!\!\!\overset{CH_2CH_2OH}{} C=CH$$

Ether (75 ml) is placed in a 1 l flask fitted with a stirrer and a pressure-equalizing dropping funnel, and furnished with an inert atmosphere. (E)-2-Ethyl-1-iodohex-1-ene (10.5 g, 50 mmol) is added and the solution is cooled to $-70°$ and maintained at that temperature and stirred as n-butyllithium (1.1 M in ether, 57 mmol)[a] is added. The mixture is stirred at $-60°$ for 15 min and then cooled to $-80°$. Stirring and cooling is continued as a solution of oxirane (4.4 g, 100 mmol) in ether (60 ml) is added rapidly.

TABLE 8.3
Ring opening of Oxiranes by Organolithium Compounds

Oxirane	Organolithium compound	Product (yield %)	Ref.
 ▲△O	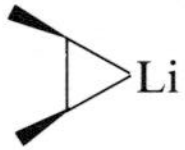 ▲△−Li	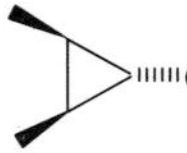 ▲△−CH$_2$CH$_2$OH (59)	(5)
△O	C$_6$Cl$_5$Li	C$_6$Cl$_5$CH$_2$CH$_2$OH (64)	(6)
△O	H$_2$C=C=C(Li)(OMe)	H$_2$C=C=C(CH$_2$CH$_2$OH)(OMe) (76)	(7)
 △O−Me	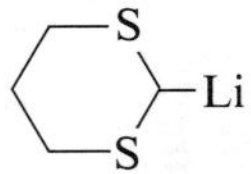 (dithiane)−Li	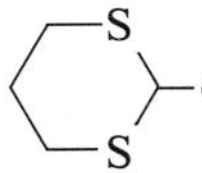 (dithiane)−CH$_2$CH(OH)Me (73)	(8)
 △O−Me	Me$_3$CCH$_2$CMe$_2$N=C(Li)(Bu)	BuCOCH$_2$CH(OH)Me^a (90)	(9)
 △O−Me	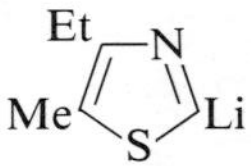 (Et,Me-thiazole)−Li	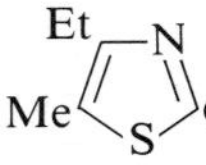 (Et,Me-thiazole)−CH$_2$CH(OH)Me (76)	(10)

TABLE 8.3—continued

Substrate	Reagent	Product (% yield)	Ref.
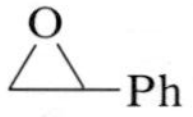 epoxide-Ph	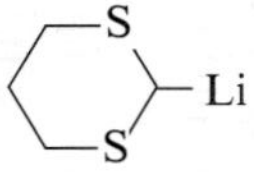 dithiane-Li	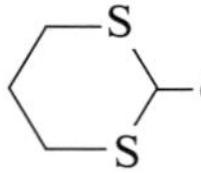 dithiane-CH₂CH(OH)Ph (87)	(11)
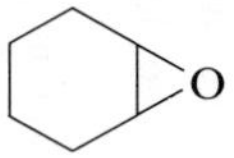 cyclohexene oxide	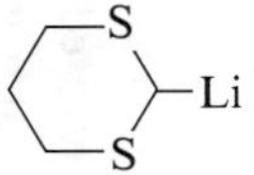 dithiane-Li	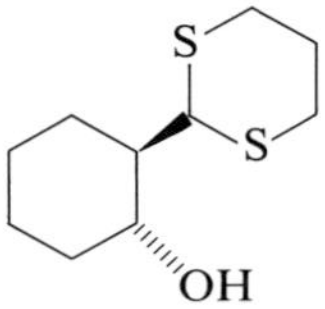 dithianyl-cyclohexanol (94)	(8)
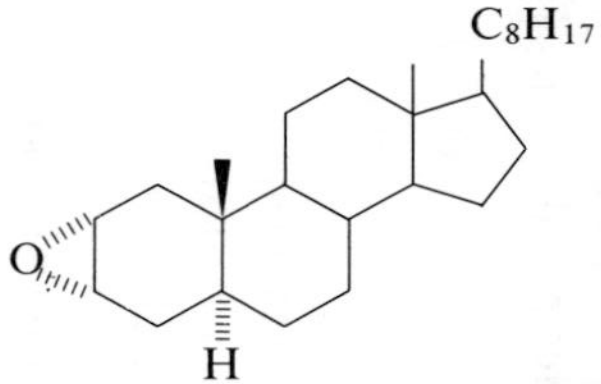 steroidal epoxide (C₈H₁₇ cholestane with α-epoxide)	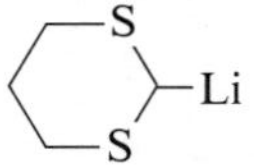 dithiane-Li	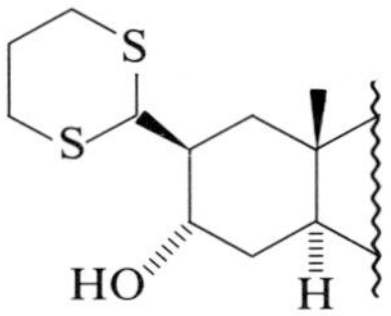 dithianyl-steroid diol (84)	(12)[b]

[a] After hydrolysis of the imine. [b] See also Ref. (13) for further examples, involving steroidal spiroepoxides.

The mixture is allowed to warm to room temperature. After 30 min, saturated aqueous ammonium chloride (80 ml) is added. The organic layer is separated. The aqueous layer is extracted with pentane (2 × 50 ml). The combined organic layers are dried (MgSO$_4$) and distilled, the fraction b.p. 109°/15 mmHg being collected as (*E*)-4-ethyloct-3-en-1-ol[b] (6.8 g, 88%).

[a] Alternatively, ethyllithium (0.98 M in ether) was used (*14*).
[b] $n_D^{20.5}$ 1.4585.

Ring opening of an epoxide promoted by boron trifluoride;
trans-*2-butylcyclohexanol (3, 15)*

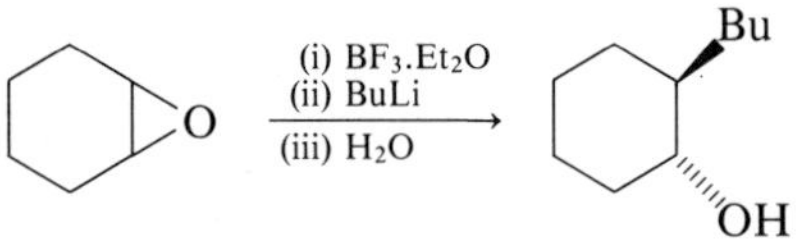

THF (7 ml) and boron trifluoride etherate (0.426 g, 3.0 mmol) are placed in a 25 ml round-bottomed flask fitted with a magnetic stirrer and a septum and furnished with an atmosphere of argon. The solution is cooled to −78° and stirred as n-butyllithium (*ca.* 2.1 M in hexane, 3.0 mmol) is added dropwise. Immediately thereafter neat cyclohexene oxide (0.098 g, 1.0 mmol) is added quickly; no colour changes are observed. Stirring at −78° is continued for 5 min, by which time TLC indicates complete consumption of the epoxide. Saturated aqueous sodium hydrogencarbonate (3 ml) is added to the cold solution. After the mixture has warmed to room temperature most of the THF is evaporated under reduced pressure. Water (3 ml) is added and the mixture is extracted with 1:1 hexane:ether (3 × 10 ml). The combined extracts are dried (MgSO$_4$) and concentrated. Flash chromatography of the residue (silica, 4:1 hexane:ethyl acetate) gives *trans*-2-butylcyclohexanol (0.151 g, 97%).

The use of oxetanes for three-carbon chain extension of organolithium compounds is comparatively little known, but gives good results with oxetane itself; examples are listed in Table 8.4:

Few similar reactions of substituted oxetanes have been reported (*16, 17, 18*); as in the case of substituted oxiranes, activation by boron trifluoride is reported to be effective in promoting the desired ring opening (*3, 19*).*

*Even tetrahydrofuran rings may be opened in the presence of boron trifluoride (*3, 20*).

8.3. NUCLEOPHILIC RING OPENING OF EPOXIDES ETC.

TABLE 8.4

Ring Opening of Oxetane by Organolithium Compounds

Organolithium compound	Product (yield %)	Ref.
PhLi	Ph(CH$_2$)$_3$OH (85)	(21)
PhCH(Li)(NC)	PhCH((CH$_2$)$_3$OH)(NC) (ca. 80)	(22)
PhCH(Li)(SiMe$_3$)	PhCH((CH$_2$)$_3$OH)(SiMe$_3$) (58)	(23)
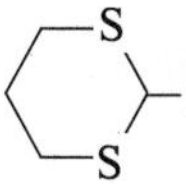 1,3-dithian-2-yl-Li	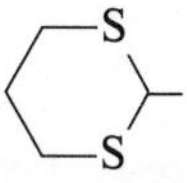 2-(CH$_2$)$_3$OH-1,3-dithiane (80)	(11)

Reaction with oxetane; 3-pentachlorophenylpropan-1-ol (7)

$$C_6Cl_6 \xrightarrow{BuLi} C_6Cl_5Li \xrightarrow[(ii)\ H_2O]{(i)\ \text{oxetane}} C_6Cl_5CH_2CH_2CH_2OH$$

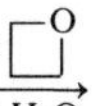

Hexachlorobenzene (5.70 g, 20 mmol) is suspended in ether (200 ml) in a three-necked flask fitted with a septum, a pressure-equalizing dropping funnel, a thermometer and a magnetic stirrer, and furnished with an atmosphere of nitrogen. The mixture is cooled to $-70°$ and stirred as n-butyllithium (ca. 2.3 M in hexane, 22 mmol) is added by syringe. The mixture is stirred at $-70°$ for 15 min and at $-20°$ for 90 min. Stirring at $-20°$ is continued as a solution of oxetane (1.16 g, 20 mmol) in ether (40 ml) is added from the dropping funnel. The mixture is stirred at room temperature for 2 h. Water (50 ml) is added. The organic layer is separated and the aqueous layer is extracted with ether. The combined organic layers are dried and evaporated. The residue is subjected to chromatography on silica. Light petroleum elutes a mixture of pentachlorobenzene and polychlorobiphenyls, and chloroform elutes 3-pentachlorophenylpropan-1-ol (2.58 g, 43%), m.p. 97–98°.

References

1. See Section 6.1.2, Ref. (*3*).
2. See Section 6.1.2, Ref. (*4*).
3. M. J. Eis, J. E. Wrobel and B. Ganem, *J. Am. Chem. Soc.* **106**, 3693 (1984).
4. N. C. Barua and R. R. Schmidt, *Synthesis* 1067 (1986).
5. M. J. S. Dewar and J. M. Harris, *J. Am. Chem. Soc.* **92**, 6557 (1970).
6. N. J. Foulger and B. J. Wakefield, *J. Chem. Soc. Perkin Trans. 1*, 871 (1974); N. J. Foulger, Ph.D. Thesis, Salford University, 1973.
7. See Section 6.4, Ref. (*14*), p. 39.
8. D. Seebach, N. R. Jones and E. J. Corey, *J. Org. Chem.* **33**, 300 (1968).
9. G. E. Niznik, W. H. Morrison and H. M. Walborsky, *J. Org. Chem.* **39**, 600 (1974).
10. J. Beraud and J. Metzger, *Bull. Soc. Chim. Fr.* 2072 (1962).
11. See Section 6.2, Ref. (*2*).
12. J. B. Jones and R. Grayshan, *Can. J. Chem.* **50**, 810 (1972).
13. J. B. Jones and R. Grayshan, *Can. J. Chem.* **50**, 1407 and 1414 (1972).
14. G. Cahiez, D. Bernard and J. F. Normant, *Synthesis* 245 (1976).
15. B. Ganem, Personal communication.
16. W. C. Still, *Tetrahedron Lett.* 2115 (1976).
17. M. Sevrin and A. Krief, *Tetrahedron Lett.* **21**, 585 (1980).
18. M. Schakel, J. J. Vrielink and G. W. Klumpp, *Tetrahedron Lett.* **28**, 5747 (1987).
19. N. C. Barua and R. R. Schmidt, *Chem. Ber.* **119**, 2066 (1986).
20. S. W. Jones, Ph.D. Thesis, Salford University, 1986.
21. S. Searles, *J. Am. Chem. Soc.* **73**, 125 (1951).
22. U. Schöllkopf, R. Jentsch, K. Madawinata and R. Harms, *Justus Liebigs Ann. Chem.* 2105 (1976).
23. A. Maercker and R. Stötzel, *Chem. Ber.* **120**, 1695 (1987).

—9—
Reactions of Organolithium Compounds with Proton Donors

9.1. FORMATION OF LITHIUM ALKOXIDES, THIOLATES AND AMIDES

The reaction of organolithium compounds with alcohols, phenols, thiols, primary and secondary amines, and the like is very general:

$$R^1Li + XH \longrightarrow R^1H + LiX$$

where $X = R^2O-, R^2S-, R^2NH-, R^2R^3N-$

It is often used as an efficient method for forming alkoxides etc. under anhydrous conditions, since the by-product R^1H is innocuous or easily removed. The example of the preparation of LDA by the reaction of n-butyllithium with diisopropylamine has already been described (see pp. 37, 108).* Other examples are listed in Table 9.1. It has also been reported that the reaction of an organolithium compound with t-butylhydroperoxide at $-78°$ gives lithium t-butylperoxide with only a little cleavage of the peroxide (2).

References

1. H. O. House, W. V. Phillips, T. S. B. Sayer and C. C. Yau, *J. Org. Chem.* **43**, 700 (1978).
2. C. Clark, P. Hermans, O. Meth-Cohn, C. Moore, H. C. Taljaard and G. van Vuuren, *J. Chem. Soc. Chem. Commun.* 1378 (1986).
3. A. J. Oliver and W. A. G. Graham, *J. Organomet. Chem.* **19**, 17 (1969).
4. E. M. Kaiser and R. A. Woodruff, *J. Org. Chem.* **35**, 1198 (1970).
5. B. A. Carlson and H. C. Brown, *Org. Synth.* **58**, 25 (1978).
6. G. H. Posner and C. E. Whitten, *Org. Synth.* **55**, 122 (1975).
7. D. Reed, D. Barr, R. E. Mulvey and R. Snaith, *J. Chem. Soc. Dalton Trans.* 557 (1986).
8. C. M. Dougherty and R. A. Olofson, *Org. Synth.* **58**, 37 (1978).
9. See Section 3.2, Ref. (*6*).
10. See Section 3.2, Ref. (*22*).

*For the preparation of a metastable solution of LDA in hydrocarbon solvent see Ref. (*1*).

9. REACTIONS WITH PROTON DONORS

TABLE 9.1
Preparation of Lithium Alkoxides, Thiolates and Amides

Proton donor	Organolithium compound	Product	Ref.
MeOH	Bu^{n}Li	MeOLi	(3)
Bu^{t}OH	Bu^{n}Li	Bu^{t}OLi[a]	(4)
Ph$_3$COH	Bu^{n}Li	Ph$_3$COLi	(5)
PhSH	Bu^{n}Li	PhSLi	(6)
C$_6$F$_5$SH	Bu^{n}Li	C$_6$F$_5$SLi	(3)
PhNH$_2$	Bu^{n}Li	PhNHLi[a,b]	(7)
C$_6$F$_5$NH$_2$	Bu^{n}Li	C$_6$F$_5$NHLi	(3)
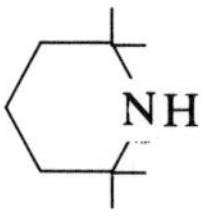 ![tetramethylpiperidine NH]	MeLi	 ![tetramethylpiperidine NLi] (LiTMP)	(8)
PhNHR (R = Me, Et)	Bu^{n}Li	PhN(Li)R	(9)
(Me$_3$Si)$_2$NH	Bu^{n}Li	(Me$_3$Si)$_2$NLi (LBTMSA)	(10)

[a] Other examples described. [b] Isolated as hexametapol complex.

9.2. HYDROGEN ISOTOPIC LABELLING VIA ORGANOLITHIUM COMPOUNDS

The reaction of an organolithium compound with an O-deuteriated or tritiated hydroxy compound is a valuable method for introducing a hydrogen isotope efficiently into a selected position:

$$\text{RLi} + \text{R'OD (T)} \longrightarrow \text{RD (T)} + \text{LiOR'}$$

The most commonly used source of the hydrogen isotope is labelled water or methanol, the latter ensuring a homogeneous solution. Labelled carboxylic acids are also occasionally used (1). Very good yields and incorporations are usually achieved; the following is an example of the use of the reaction with D$_2$O to accomplish multiple labelling of benzoic acid derivatives (2):

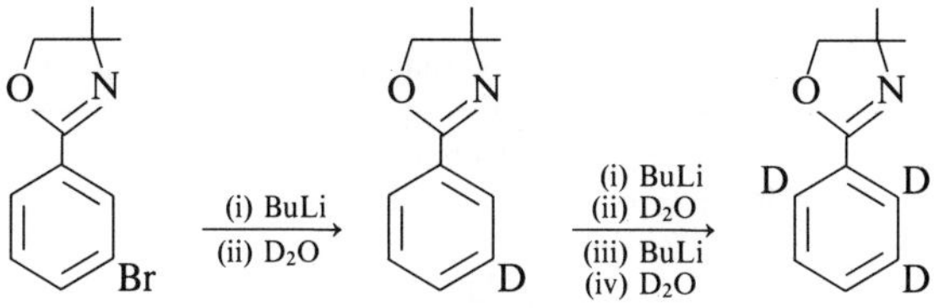

9.2. HYDROGEN ISOTOPIC LABELLING

[Scheme: aryloxazoline-Br → (i) BuLi, (ii) D₂O → aryloxazoline-D → (i) BuLi, (ii) D₂O, (iii) BuLi, (iv) D₂O → tetradeuterated aryloxazoline]

74%

($d_0 : d_1 : d_2 : d_3 : d_4 =$ 0 : 4 : 44 : 100 : 11)

However, the reaction is not always as free from complications as is commonly believed. This is because the rate of deuteriation (tritiation) is not always much faster than that of competing reactions (3–5). The very high rate of a lithium–halogen exchange reaction has in fact been taken advantage of: reactions of n-butyllithium with aryl halides *in the presence* of tritiated water led to tritium labelling of the aromatic ring (6). Nevertheless, this remains one of the best methods for hydrogen isotopic labelling in specific positions and with very high incorporations. Some examples of deuteriation are shown in Table 9.2. Some earlier examples are listed in General Ref. A, which also gives references to tritium labelling. A more recent example of the latter is shown below (7):

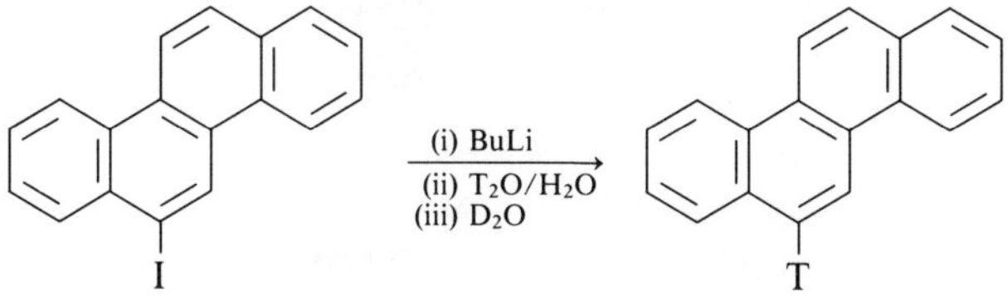

[Scheme: chrysene-I → (i) BuLi, (ii) T₂O/H₂O, (iii) D₂O → chrysene-T]

70% chemical yield
28 mCi mmol^{-1}

N-*(1-d-2-Methylbutylidene)-1,1,3,3-tetramethylbutylamine* and *1-d-2-methylbutanal* (13)

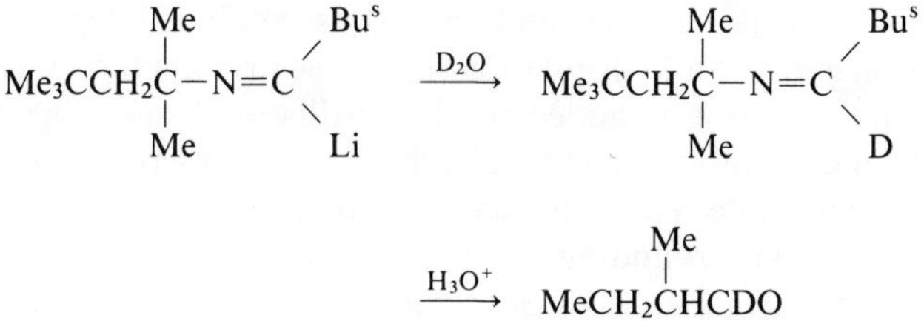

$$\text{Me}_3\text{CCH}_2\overset{\text{Me}}{\underset{\text{Me}}{\text{C}}}-\text{N}=\text{C}\overset{\text{Bu}^s}{\underset{\text{Li}}{\diagup\diagdown}} \xrightarrow{\text{D}_2\text{O}} \text{Me}_3\text{CCH}_2\overset{\text{Me}}{\underset{\text{Me}}{\text{C}}}-\text{N}=\text{C}\overset{\text{Bu}^s}{\underset{\text{D}}{\diagup\diagdown}}$$

$$\xrightarrow{\text{H}_3\text{O}^+} \text{MeCH}_2\overset{\text{Me}}{\underset{}{\text{C}}}\text{HCDO}$$

TABLE 9.2

Deuterium Labelling via Organolithium Compounds

Organolithium compound	Source of deuterium	Labelled product (yield %)	Incorporation (%)	Ref.
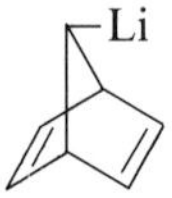 norbornenyl-Li	D_2O	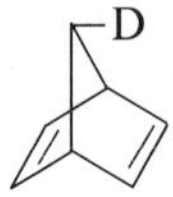 norbornenyl-D (39)[a]	96 ± 1	(8)
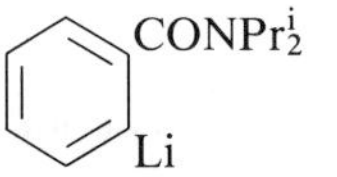 2-Li-C$_6$H$_4$-CONPr$^i_2$	D_2O	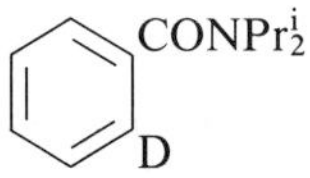 2-D-C$_6$H$_4$-CONPr$^i_2$ (90)	97	(9)
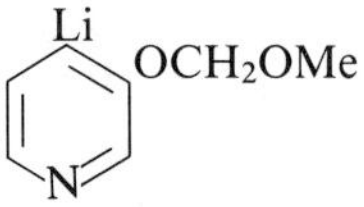 3-Li-2-(OCH$_2$OMe)pyridine	D_2O	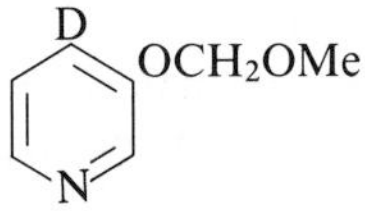 3-D-2-(OCH$_2$OMe)pyridine (88)	95	(10)
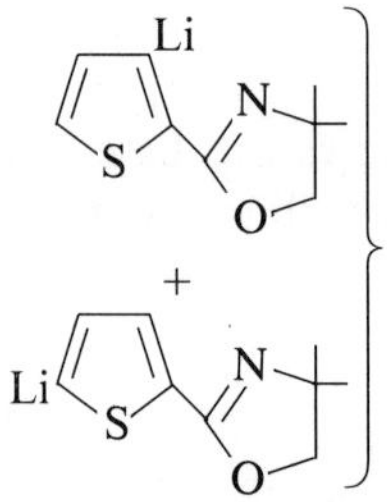 2-Li-thienyl-oxazoline + 5-Li-thienyl-oxazoline	MeOD	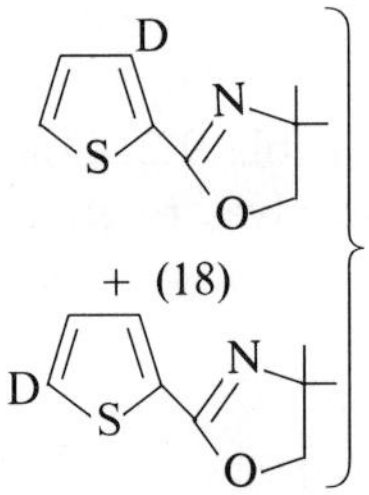 2-D-thienyl-oxazoline (82) + 5-D-thienyl-oxazoline (18) (100)	100	(11)
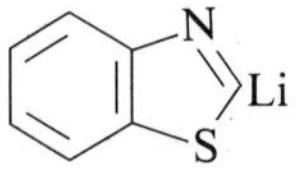 2-Li-benzothiazole	D_2O	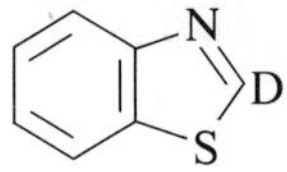 2-D-benzothiazole (95)	98	(12)

[a] The low yield is a consequence of the rigorous isolation and purification procedure used.

A solution of 4-lithio-3,6,6,8,8-pentamethyl-5-azanon-4-ene is prepared as described on p. 66 and stirred rapidly at *ca.* −5° as deuterium oxide (8.0 ml, 0.40 mol) is injected rapidly; the temperature of the solution rises to *ca.* 30°. The cooling bath is removed and stirring is continued for 30 min. The solution is filtered into a 1 l distillation flask. The reaction flask is rinsed with pentane, and the rinse is added to the distillation flask. The solvent is evaporated and the residue fractionally distilled, the fraction b.p. 52.5–54°/1.5 mmHg being collected as the deuterioaldimine (33.7–34.9 g, 85–88%).

A 1 l three-necked round-bottomed flask is equipped with a dropping funnel, a steam inlet tube, and a Dean and Stark trap fitted with a condenser through which acetone cooled to $-15°$ is circulated. A solution of oxalic acid dihydrate (50.4 g, 0.40 mol) in water (200 ml) is added to the flask and heated to reflux. Steam is passed into the flask, and when some begins to condense in the trap the deuterioaldimine is added dropwise from the funnel. The aldehyde and water collect in the trap, from which the water is periodically removed. After all the aldehyde has distilled, the distillate is drained from the trap. The water layer is separated and discarded. The organic layer is washed with saturated aqueous sodium chloride (3 × 25 ml), dried ($CaSO_4$) and fractionally distilled to give 1-d-2-methylbutanal (13.0–13.3 g, 87–88%), b.p. 92–93°, with an isotopic purity of $ca.$ 98%.

References

1. L. Lompa-Krzymien and L. C. Leitch, *Synthesis* 124 (1976).
2. A. T. Meyers and E. D. Mihelich, *J. Org. Chem.* **40**, 3158 (1975).
3. P. Beak and C.-W. Chen, *Tetrahedron Lett.* **26**, 4979 (1985).
4. N. S. Narasimhan and R. Ammanamanchi, *J. Chem. Soc. Chem. Commun.* 1368 (1985).
5. See also R. A. Firestone and B. J. Christensen, *J. Chem. Soc. Chem. Commun.* 288 (1976); P. J. Crowley, M. R. Leach, O. Meth-Cohn and B. J. Wakefield, *Tetrahedron Lett.* **27**, 2909 (1986).
6. R. Taylor, *Tetrahedron Lett.* 435 (1975).
7. G. M. Blackburn, A. J. Flavell, L. Orgee, J. P. Will and G. M. Williams, *J. Chem. Soc. Perkin Trans. 1*, 3196 (1981).
8. See Section 3.1.2, Ref. (*7*).
9. See Section 3.2, Ref. (*37*).
10. See Section 3.2, Ref. (*11*).
11. See Section 3.2, Ref. (*18*).
12. H. Chikashita and K. Itoh, *Heterocycles* **23**, 293 (1985).
13. See Section 5.4, Ref. (*2*).

9.3. INDIRECT REDUCTION OF ORGANIC HALIDES

The following reaction sequence represents an overall reduction of the organic halide:

$$RX \xrightarrow{R'Li \text{ or } Li} RLi \xrightarrow{H^+} RH$$

It has not been extensively applied, but can be useful because of its selectivity and/or because of the mild conditions required (especially with metal–halogen exchange as the first step). For example, reaction of the dibromocyclopropane [1] with methyllithium at −80° followed by hydrolysis gave exclusively the *exo*-monobromocyclopropane [2] in high yield (*1*):*

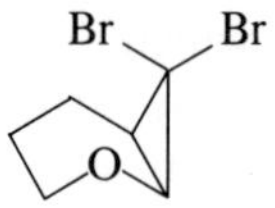

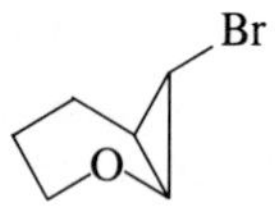

Reference

1. K. G. Taylor, W. E. Hobbs and M. Saquet, *J. Org. Chem.* **36**, 369 (1971).

*And deuterolysis gave 7-*exo*-bromo-7-*endo*-deuterio-2-oxabicyclo[4.1.0]heptane.

—10—

Formation of Carbon–Nitrogen Bonds via Organolithium Compounds

Much effort has been devoted to achieving conversion of organolithium compounds into compounds with carbon–nitrogen bonds in place of carbon–lithium bonds; a good synthesis of amines has been particularly sought. Most of the reported reactions involve the displacement of a good leaving group from nitrogen, or addition to nitrogen–nitrogen multiple bonds (see General Ref. D(ii)). A selection is shown below:

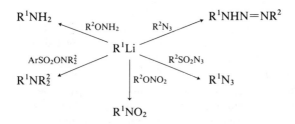

Of these methods, the most generally useful are those with hydroxylamine derivatives and those with azides.

10.1. REACTIONS WITH HYDROXYLAMINE DERIVATIVES

Various O-substituted hydroxylamines have been tried for converting organolithium compounds into the corresponding primary amines. The most consistently useful of them, methoxyamine, suffered from the disadvantage that two or more equivalents of the organolithium compound were required. This disadvantage was overcome when Beak confirmed that a methoxyamine–methyllithium reagent could be used (1, 2). It was subsequently shown that secondary amines could be prepared similarly from N-substituted methoxyamines (2–4). Boche et al. have successfully employed various *O*-phosphorus derivatives of hydroxylamine and *N*-substituted hydroxylamines (5) and have observed significant asymmetric induction by using the reagent [1] derived from (−)-ephedrine (6):

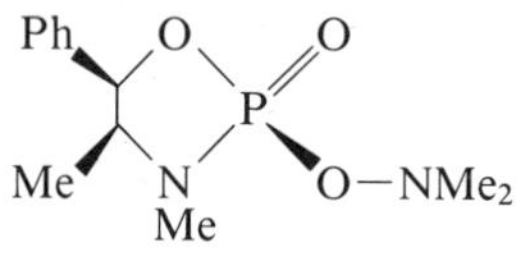

Ph O   O
 \ / \ //
  C   P
 / \ / \
Me N   O—NMe$_2$
    |
    Me

[1]

Amination of phenyllithium (1)

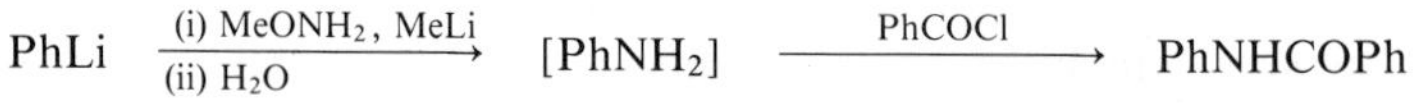

PhLi $\xrightarrow[\text{(ii) H}_2\text{O}]{\text{(i) MeONH}_2, \text{MeLi}}$ [PhNH$_2$] $\xrightarrow{\text{PhCOCl}}$ PhNHCOPh

Methyllithium[a] (*ca.* M in ether, 8.5 mmol) is stirred and cooled to −78° under an atmosphere of nitrogen. A solution of methoxyamine (0.40 g, 8.5 mmol) in hexane (9 ml) is added at a rate of approximately 1 drop s^{-1}, followed by phenyllithium (*ca.* 1.5 M in 30:70 ether–cyclohexane, 4.3 mmol). The mixture is allowed to warm to −15° and held at that temperature for 2 h. Water (0.5 ml) is added, followed by a mixture of pyridine (7 ml) and ether (6 ml) and then by a solution of benzoyl chloride (1.9 ml) in ether (7 ml). The mixture is stirred overnight. The product is isolated by extraction with chloroform and purified by chromatography (silica, 20% ethyl acetate–hexane) to give *N*-phenylbenzamide (0.76 g, 90%), m.p. 161.5–163°.

[a] The presence or absence of lithium bromide is immaterial.

References

1. P. Beak and B. J. Kokko, *J. Org. Chem.* **47**, 2823 (1982).
2. P. Beak, A. Basha, B. Kokko and D. Loo, *J. Am. Chem. Soc.* **108**, 6106 (1986).
3. B. J. Kokko and P. Beak, *Tetrahedron Lett.* **24**, 561 (1983).
4. P. Beak, A. Basha and B. J. Kokko, *J. Am. Chem. Soc.* **106**, 1511 (1984); see also G. Boche and H.-U. Wagner, *J. Chem. Soc. Chem. Commun.* 1591 (1984).
5. G. Boche, M. Bernheim and W. Schrott, *Tetrahedron Lett.* **23**, 5399 (1982).
6. G. Boche and W. Schrott, *Tetrahedron Lett.* **23**, 5403 (1982).

10.2. REACTIONS WITH AZIDES

The course of reactions of organolithium compounds with azides is as follows:

$$RLi + X-\overset{\ominus}{N}-N=N^{\oplus} \longrightarrow X-\underset{Li}{N}-N-N-R \underset{-LiX}{\overset{H_3O^+}{\diagup\diagdown}} \begin{matrix} X-NHN=NR \\ \text{or tautomer} \\ \\ RN_3 \end{matrix}$$

If the group X is a poor leaving group then hydrolysis gives a triazene (and further hydrolysis leads to the amine) (1). However, if X is a good leaving group such as arenesulphonyl then elimination gives the azide corresponding to the organolithium compound. As in the following example, the azide can be isolated or it can be reduced to the corresponding amine. Other examples are described in Refs (2)–(4).

Alternative azides for these types of reaction are diphenyl phosphorazidate (5) and trimethylsilylmethyl azide (6); both are reported to be less hazardous than tosyl azide, though on the basis of limited experience.

Reaction with a sulphonyl azide; 2-azido-3,3'-bithienyl and 2-amino-3,3'-bithienyl (7)[a]

$$\underset{Br}{\text{S}\diagup\diagdown\text{S}} \xrightarrow[\text{(ii) 4-MeC}_6\text{H}_4\text{SO}_2\text{N}_3]{\text{(i) Bu}^n\text{Li}} \underset{N_3}{\text{S}\diagup\diagdown\text{S}} \xrightarrow{\text{LiAlH}_4} \underset{NH_2}{\text{S}\diagup\diagdown\text{S}}$$

n-Butyllithium (ca. 1.6 M in n-hexane, 51 mmol) is cooled to −70° under an atmosphere of nitrogen. A solution of 2-bromo-3,3'-bithienyl (12.2 g, 50 mmol) in ether (50 ml) is added dropwise with stirring at −70° and the mixture is stirred for a further 30 min at that temperature. Cooling to −70° is continued as an ethereal solution of p-toluenesulphonyl azide (10 g, 55 mmol) is added dropwise, and for a further 5 h. The temperature of the mixture is allowed to rise to −10° and the suspension is rapidly filtered. The precipitate is washed with dry ether and then suspended in n-pentane (150 ml) at −70°. A solution of tetrasodium pyrophosphate (13.3 g, 50 mmol) in water (200 ml) is added. When the temperature of the mixture has reached 0°, it is stirred for a few minutes, then cooled to −20° and kept at this temperature overnight. The organic phase is separated and dried, and the solvent is evaporated under reduced pressure at low temperature. The residual 2-azido-3,3'-bithienyl (41%), a yellow solid, is stored at −20°.

Lithium aluminium hydride (1.0 g, 25 mmol)[b] is suspended in dry ether (30 ml). The suspension is stirred at 0° as a cold ($-20°$) solution of the azide (1.0 g, 5 mmol) in ether (20 ml) is added dropwise. The mixture is stirred at $-20°$ for 2 h and at room temperature for 2 h. It is again cooled as wet ether is added, followed by cold distilled water. The mixture is filtered and the filtrate is extracted with ether. The combined extracts are washed with water and dried, and the solvent is evaporated under reduced pressure. Distillation of the residue gives 2-amino-3,3'-bithienyl (87%), b.p. 120–122°/0.5 mmHg.

[a] Several other examples are described.
[b] Hydrogen sulphide (7) or sodium borohydride (2) may also be used for the reduction.

References

1. L. I. Skripnik and V. Ya. Pochinok, *Chem. Heterocycl. Compds (Engl. Transl.)* 729 (1968); K. L. Kirk, *J. Org. Chem.* **43**, 4381 (1978); see also B. M. Trost and W. H. Pearson, *J. Am. Chem. Soc.* **103**, 2483 (1981) and **105**, 1054 (1983).
2. J. N. Reed and V. Snieckus, *Tetrahedron Lett.* **24**, 3795 (1983) and **25**, 5505 (1984).
3. Y. Kawada, H. Yamazaki, G. Koga, S. Murata and H. Iwamura, *J. Org. Chem.* **51**, 1472 (1986).
4. N. S. Narasimhan and R. Ammanamanchi, *Tetrahedron Lett.* **23**, 4733 (1983).
5. S. Mori, T. Aoyama and T. Shioiri, *Tetrahedron Lett.* **27**, 6111 (1986).
6. T. R. Kelly and M. P. Maguire, *Tetrahedron* **41**, 3033 (1985).
7. P. Spagnolo, P. Zanirato and S. Gronowitz, *J. Org. Chem.* **47**, 3177 (1982).

—11—
Formation of Carbon–Oxygen Bonds via Organolithium Compounds

The functional-group interconversion RLi → ROH or ROX could be very useful, but only a limited number of methods is available, and none is routinely satisfactory. Of the "direct" methods, reactions with dioxygen and peroxides are the best established. Reaction with the molybdenum pentoxide–pyridine–hexametapol complex ("MoOPH") is relatively untried, but promising.

11.1. REACTION WITH DIOXYGEN

The reaction of organolithium compounds with dioxygen can be controlled to give at low temperatures a hydroperoxide or at higher temperatures an alcohol or phenol.

$$RLi + O_2 \longrightarrow ROOLi \xrightarrow{RLi} 2ROLi$$
$$\downarrow H_3O^+ \qquad\qquad \downarrow H_3O^+$$
$$ROOH \qquad\qquad 2ROH$$

However, many side-reactions occur and the yields of the desired products are often poor: in a recent study yields of phenols from aryllithium compounds ranged from 34 to 52% (1). Other examples are listed in General Ref. A. An exception to this generalization is the oxidation of lithium enolates, which can give good yields of either hydroxy or hydroperoxy derivatives, as in the following example. For further examples of reactions of organolithium compounds with dioxygen see Refs (2)–(4).

Yields are usually improved when the organolithium compound is converted into an organomagnesium compound *in situ* by the addition of magnesium bromide, or when it is co-oxidized with an expendable Grignard reagent (see General Refs A and D(ii)) as in the following example (5):

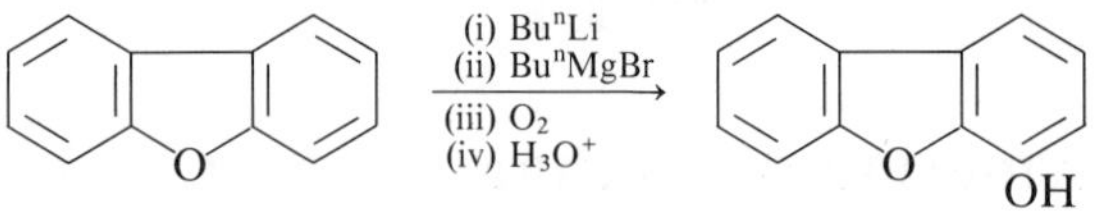

dibenzofuran $\xrightarrow[\text{(iv) H}_3\text{O}^+]{\text{(i) Bu}^n\text{Li} \quad \text{(ii) Bu}^n\text{MgBr} \quad \text{(iii) O}_2}$ 4-hydroxydibenzofuran

50–60%

An example of a related indirect procedure, involving the preparation and oxidation of an organoboron intermediate, is described on p. 151.*

Reaction with dioxygen; α-hydroxyphenylacetic acid and α-hydroperoxyphenylacetic acid (7)

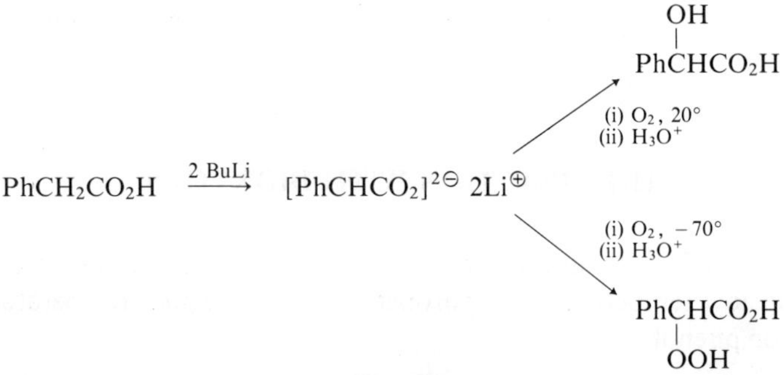

$$PhCH_2CO_2H \xrightarrow{2\ BuLi} [PhCHCO_2]^{2\ominus}\ 2Li^{\oplus}$$

$(i)\ O_2,\ 20°$
$(ii)\ H_3O^+$ → $\underset{|}{\overset{OH}{PhCHCO_2H}}$

$(i)\ O_2,\ -70°$
$(ii)\ H_3O^+$ → $\underset{|}{\overset{}{PhCHCO_2H}}\ \text{OOH}$

A 100 ml round-bottomed two-necked flask is fitted with a magnetic stirrer and a rubber septum, and furnished with a nitrogen atmosphere. A solution of phenylacetic acid (0.17 g, 1.4 mmol) in THF (30 ml) is introduced by syringe. The solution is cooled to $-60°$ and stirred while n-butyllithium (ca. 2 M in hexane, 1.4 mmol) is added dropwise by syringe during ca. 5 min. After 60 min the mixture is allowed to warm to $-40°$ and further n-butyllithium (1.4 mmol) is added dropwise. During this stage the colour of the solution turns from pale-yellow to dark-red. The mixture is stirred at $-40°$ for a further 100 min.

(a) The solution is allowed to warm to ca. 20° and stirred as a fast stream of dry oxygen is passed into it by means of a stainless-steel capillary tube. When the resulting precipitation is complete, the solvent is removed by means of a rotary evaporator (ca. 30°/20 mmHg). The residue is treated with 10%

---

* For a one-step modification of this procedure see Ref. (6).

hydrochloric acid and extracted with ether (4 × 5 ml). The combined extracts are dried ($MgSO_4$) and the solvent is evaporated, leaving α-hydroxyphenylacetic acid (86%).

(b) A 100 ml round-bottomed two-necked flask is fitted with a magnetic stirrer and a septum, and furnished with a nitrogen atmosphere. THF (50 ml) is introduced, cooled to −78°, and saturated with dry oxygen, introduced through the septum by means of a stainless-steel capillary inlet. The solution is stirred and maintained at −70°, with continuous bubbling of oxygen, as a solution of the dilithiocarboxylate, prepared as described above, is added dropwise during 110 min by means of a stainless-steel capillary syphon, passed through the septum. The mixture is stirred at −78° for 30 min, and this temperature is rigorously maintained as 10% hydrochloric acid is added dropwise by means of a syringe. The mixture is allowed to warm to 5°, transferred to a separating funnel, and diluted with approximately two volumes of ice. It is extracted with ether (6 × 10 ml) while the temperature is kept below 10° by the addition of more ice. The combined extracts are dried ($MgSO_4$) in a refrigerator. First the ether is evaporated in a rotary evaporator (10°/10 mmHg), followed by the remaining solvent (10°/0.1 mmHg). The residue is crude α-hydroperoxyphenylacetic acid (82%).[a]

[a] Three recrystallizations from ether–benzene gave a sample, m.p. 96–97° dec, 96% pure by iodometric titration (7).

References

1. K. A. Parker and K. A. Koziski, *J. Org. Chem.* **52**, 674 (1987).
2. See Section 3.1.3, Ref. (*15*).
3. T. Cuvigny, G. Valette, M. Larcheveque and H. Normant, *J. Organomet. Chem.* **155**, 147 (1978).
4. W. Meyer and D. Seebach, *Chem. Ber.* **113**, 1304 (1980).
5. See Section 3.2, Ref. (*55*).
6. R. W. Hoffman and K. Ditrich, *Synthesis* 107 (1983).
7. W. Adam and O. Cueto, *J. Org. Chem.* **42**, 38 (1977).

11.2. REACTION WITH PEROXIDES

The general reaction of an organolithium compound with a peroxide is as follows, where the groups R^2 and R^3 can be alkyl, aryl, acyl or a metal:

$$R^1Li + R^2OOR^3 \longrightarrow R^1OR^2 + LiOR^3$$

TABLE 11.1

Reactions of Organolithium Compounds with Bis(trimethylsilyl)peroxide[a]

Organolithium compound	Silyl ether (yield %)	Hydrolysis product (overall yield from organolithium compound, %)	Ref.
BuLi	—	BuOH (98)	(2)
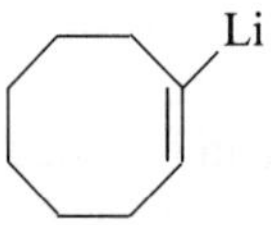 cyclooctenyl-Li	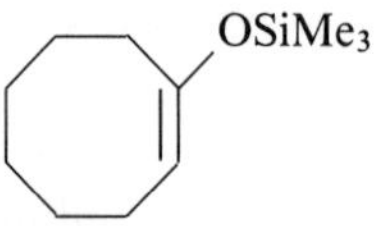 cyclooctenyl-OSiMe₃ (52)	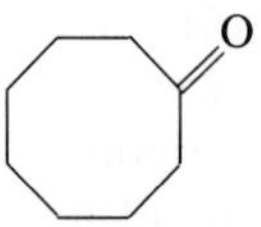 cyclooctenone (59)[b]	(1)
PhLi	PhOSiMe₃ (48)[c]	PhOH (86)	(1, 2)
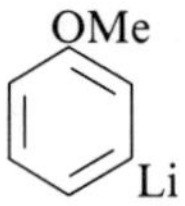 o-MeO-C₆H₄-Li	—	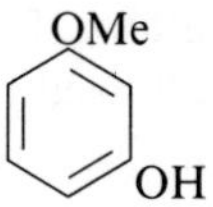 o-MeO-C₆H₄-OH (88)	(2)
 2-thienyl-Li	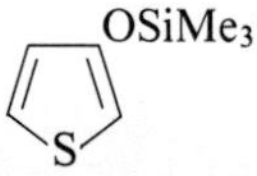 2-thienyl-OSiMe₃ (60)	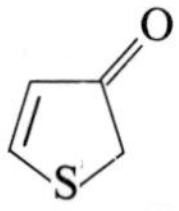 2(5H)-thiophenone (65)	(2, 3)

[a] Refs (1) and (2) give good experimental details for the preparation of the peroxide. [b] As 2,4-dinitrophenylhydrazone. [c] Mixture with phenol.

Examples of several variants on this type of reaction have been described (see General Ref. D(ii) and Ref. (1)). Probably the most generally useful is the reaction with bis(trimethylsilyl)peroxide, giving trimethylsilyl ethers, which may be easily hydrolysed to the corresponding hydroxy-compounds (1–3). Several examples are given in each of the references cited; some representative examples are listed in Table 11.1.

References

1. See Section 3.1.3, Ref. (15).
2. M. Taddei and A. Ricci, *Synthesis* 633 (1986).
3. L. Camici, A. Ricci and M Taddei, *Tetrahedron Lett.* **27**, 5155 (1986).

11.3. REACTION WITH MOLYBDENUM PENTOXIDE–PYRIDINE–HEXAMETAPOL

Molybdenum pentoxide forms a 1:1:1 complex with pyridine and hexametapol (MoOPH), which is readily prepared,* and under the recommended storage conditions is stable for several months (*1*). In the course of a study of reactions of n-butyllithium with transition-metal peroxides, Regen and Whitesides observed that a molybdenum pentoxide–hexametapol complex gave lithium butoxide (*3*). The MoOPH reagent has subsequently been used to oxidize various organolithium compounds, but it has been found particularly effective with lithium enolates, including some that give poor results on direct oxygenation. The α-hydroxylation of camphor has been fully described, and conditions for hydroxylation of other ketones recommended (*1*). Besides carbonyl compounds (*1*, *4*), types of compound that have been hydroxylated by lithiation followed by reaction with MoOPH include nitriles (*2*) and sulphones (*6*); in the latter case elimination gives a ketone:

$$R^1R^2CHSO_2Ar \xrightarrow[\text{(ii) MoOPH}]{\text{(i) LDA}} R^1R^2C\diagup^{SO_2Ar}_{\diagdown OLi} \xrightarrow{-LiSO_2Ar} R^1R^2C=O$$

References

1. E. Vedejs and S. Larsen, *Org. Synth.* **64**, 127 (1986).
2. A. R. Daniewski and W. Wojciechowska, *Synth. Commun.* **16**, 535 (1986).
3. S. L. Regen and G. M. Whitesides, *J. Organomet. Chem.* **59**, 293 (1973).
4. E. Vedejs, D. A. Engler and J. E. Telschow, *J. Org. Chem.* **43**, 188 (1978); H. Niwa, T. Hasegawa, N. Ban and K. Yamada, *Tetrahedron* **43**, 825 (1987).
5. E. Vedejs and J. E. Telschow, *J. Org. Chem.* **41**, 740 (1976).
6. R. D. Little and S. O. Myong, *Tetrahedron Lett.* **21**, 3339 (1980).

---

*The *Organic Syntheses* procedure involves oxidation of molybdenum trioxide by hydrogen peroxide (*1*). It has been reported that molybdic acid monohydrate can be used in place of molybdenum trioxide (*2*).

—12—
Formation of Carbon–Sulphur Bonds via Organolithium Compounds*

In contrast with the paucity of reactions for forming carbon–nitrogen bonds and carbon–oxygen bonds, there are many methods for forming carbon–sulphur bonds via organolithium compounds. One of them, thiophilic addition to thiocarbonyl groups, has been covered in Section 7.2. Of the others, reactions with elemental sulphur and cleavage of disulphides are very useful and general. Reactions with sulphur halides (and the related reactions with sulphenyl halides and thiocyanates) have been less commonly used. Of the various reactions with compounds containing sulphur–oxygen double bonds, that with sulphur dioxide is particularly useful. Thiophilic cleavage of thioethers is only valuable in special cases (see General Ref. D(ii)).

12.1. REACTION WITH ELEMENTAL SULPHUR

The course of a reaction of an organolithium compound with elemental sulphur is as follows:

$$RLi + S_n \longrightarrow RS_nLi \xrightarrow{RLi} RS_{n-m}R + Li_2S_m$$

$$\downarrow RLi$$

$$RS_{n-m-p}R + LiS_pR \text{ etc.}$$

Under carefully controlled conditions dialkyl polysulphides may be obtained (1, 2), but normally the final product is mainly a lithium thiolate. The lithium thiolate may be protonated to give the thiol, as in the detailed example below, or caused to react with other electrophiles to give, for example, thioethers or thiolesters. Some other examples are listed in Table 12.1.

*Several of the reactions described below have also been applied to the synthesis of organoselenium compounds, and examples are noted where appropriate.

TABLE 12.1
Reactions of Organolithium Compounds with Sulphur

Organolithium compound	Reagent for thiolate	Product (yield %)	Ref.
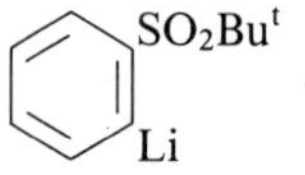 C₆H₄(SO₂Buᵗ)(Li) [2-lithio-tert-butylsulfonylbenzene]	H⁺	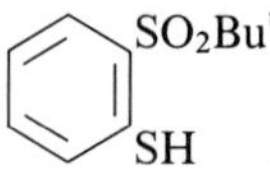 2-(SO₂Buᵗ)C₆H₄SH (85)	(3)
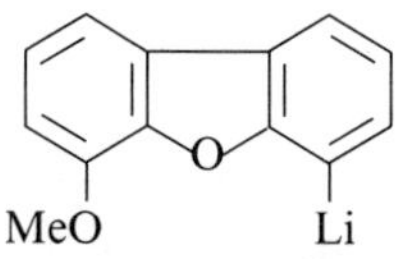 4-MeO, 6-Li dibenzofuran	H⁺	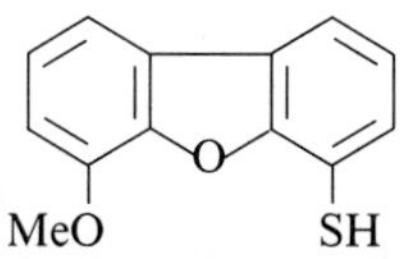 4-MeO, 6-SH dibenzofuran (43)	(4)
2-lithiofuran	Ac₂O	2-(SCOMe)furan (45)	(5)ᵃ
MeC≡CLi	ClCN	MeC≡CSCN (45)	(6)
	MeSSO₂Me	MeC≡CSSMe (54–68)	(7)
BuᵗC≡CLi	Me₃SiCl	BuᵗC≡CSSiMe₃ (60)	(8)

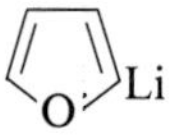

ᵃ Similarly with selenium.

Thiophene-2-thiol (9)

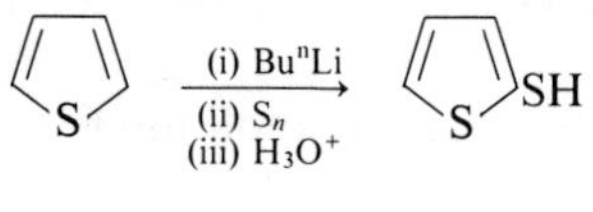

$$\text{thiophene} \xrightarrow[\text{(ii) } S_n]{\text{(i) Bu}^n\text{Li}} \xrightarrow{\text{(iii) H}_3\text{O}^+} \text{2-SH-thiophene}$$

A 3 l flask is fitted with a stirrer and a 600 ml dropping funnel and furnished with a nitrogen atmosphere. Thiophene (56.5 g, 53 ml, 0.67 mol) and THF (500 ml) are placed in the flask and stirred and cooled to −40°. The temperature is maintained at *ca.* −40° while n-butyllithium (*ca.* 1.4 M in pentane, 0.66 mol) is added from the dropping funnel during 5 min. The temperature of the mixture is held between −30° and −20° for 1 h and then lowered to −70°. Powdered sulphur crystals (20.4 g, 0.67 mol) are added in one portion to the stirred mixture. After 30 min the temperature is allowed to rise to −10°. The solution is poured into rapidly stirred ice–water (1 l). The organic layer is separated and extracted with water (3 × 100 ml). The extracts are combined with the aqueous layer, cooled, and acidified with 4 M sulphuric acid. The acidified mixture is immediately extracted with ether (3 × 200 ml). The combined extracts are washed with water (2 × 100 ml) and dried

12.2. REACTION WITH DISULPHIDES

(Na_2SO_4). The ether is distilled, and the residue is distilled under reduced pressure, the fraction b.p. 53–56°/5 mmHg being collected as thiophene-2-thiol (49.5–53.5 g, 65–70%), n_D^{25} 1.6110.

References

1. J. J. Boscato, J. M. Catala, E. Franta and J. Brossas, *Tetrahedron Lett.* **21**, 1519 (1980); *Makromol. Chem.* **180**, 1571 (1979).
2. J. J. Bishop, A. Davison, M. L. Katcher, D. W. Lichtenberger, R. E. Merrill and J. C. Smart, *J. Organomet. Chem.* **27**, 241 (1971).
3. Ya, L. Gol'dfarb, F. M. Stoyanovich, G. B. Chermanova and E. D. Lubuzh, *Izv. Akad. Nauk. SSSR Ser. Khim.* 2760 (1978).
4. See Section 3.2, Ref. (*55*).
5. E. Niwa, H. Aoki, H. Tanaka, K. Munakata and M. Namiki, *Chem. Ber.* **99**, 3215 (1966).
6. See Section 6.4, Ref. (*14*), p. 67.
7. See Section 6.4, Ref. (*14*), p. 69.
8. R. S. Sukhai, J. Meijer and L. Brandsma, *Recl. Trav. Chim. Pays-Bas* **96**, 179 (1977).
9. E. Jones and I. M. Moodie, *Org. Synth.* **50**, 104 (1970).

12.2. REACTION WITH DISULPHIDES

The cleavage of disulphides by organolithium compounds is so efficient and general that it is often used to characterize organolithium compounds; dimethyl and diphenyl disulphide are most commonly employed (see e.g. the tables in Ref. (*1*)):

$$R^1Li + R^2SSR^2 \longrightarrow R^1SR^2 + LiSR^2$$

Some well-described examples are listed in Table 12.2. The following detailed example illustrates the mild conditions required.

E-1-Phenylthio-1-hexene (11)

A solution of *E*-1-hexenyllithium is prepared as described on p. 29 from *E*-1-bromo-1-hexene (0.98 g, 6 mmol) in Trapp mixture at −120°. The temperature is maintained at −80° as a solution of diphenyldisulphide (1.31 g, 6 mmol) in THF (10 ml) is added. The solution is stirred at −70° for 10 min and at room temperature for 45 min. The reaction mixture is shaken with a

TABLE 12.2

Synthesis of Thioethers from Organolithium Compounds and Disulphides

Organolithium compound	Disulphide	Product (yield %)	Ref.
n-$C_{14}H_{29}$CH(Li)CO_2Li	MeSSMe	n-$C_{14}H_{29}$CH(SMe)(CO_2H) (90)	(2)
H_2C=C=C(OMe)(Li)	MeSSMe	H_2C=C=C(OMe)(SMe) (76)	(3)
 N-methylpyrrole-Li	MeSSMe	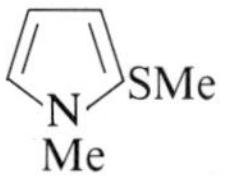 N-methylpyrrole-SMe (68)	(4)
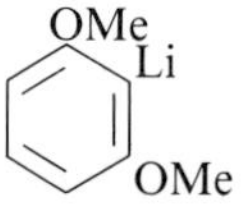 2,6-dimethoxyphenyl-Li	EtSSEt	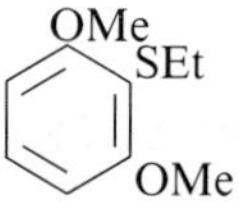 2,6-dimethoxyphenyl-SEt (83)	(5)
 2-(SPr^i)phenyl-Li	Pr^iSSPr^i	 1,2-bis(SPr^i)benzene (68)	(6)
(EtO)$_2$P(O)CH(Me)(Li)	PhSSPh[a]	(EtO)$_2$P(O)CH(Me)(SPh) (87)	(7)
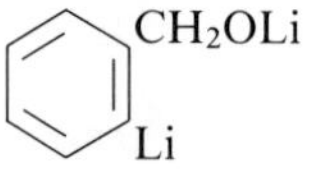 2-(CH_2OLi)phenyl-Li	PhSSPh	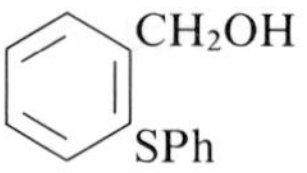 2-(CH_2OH)phenyl-SPh (72)	(8)
 2-(CH_2Li)-3-OMe-benzoate (COOEt)	PhSSPh	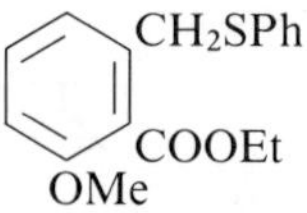 2-(CH_2SPh)-3-OMe-benzoate (COOEt) (84)	(9)

[a] Similarly with PhSeSePh; for analogous syntheses of selenoferrocenes see Ref. (10).

mixture of dichloromethane and saturated aqueous sodium chloride. The aqueous phase is separated and extracted three times with dichloromethane. The combined organic phases are extracted with 7% aqueous potassium hydroxide, washed with saturated aqueous sodium chloride, and dried

(Na_2SO_4). The solvents are evaporated in a rotary evaporator. Distillation of the residue gives E-1-phenylthio-1-hexene (1.07 g, 93%), b.p. 110–115°/ 0.8 mmHg, n_D^{20} 1.5548.

References

1. See Section 3.2, Ref. (*3*).
2. B. M. Trost and Y. Tamaru, *J. Am. Chem. Soc.* **99**, 3101 (1977).
3. See Section 6.4, Ref. (*14*), p. 42.
4. S. Gronowitz and R. Kada, *J. Heterocycl. Chem.* **21**, 1041 (1984).
5. P. Jacob and A. T. Shulgin, *Synth. Commun.* **11**, 957 (1981).
6. See Section 3.2, Ref. (*14*).
7. F. M. Hauser, R. P. Rhee, S. Prasanna, S. M. Weinreb and J. H. Dodd, *Synthesis* **72** (1980).
8. W. Meyer and D. Seebach, *Chem. Ber.* **133**, 1304 (1980).
9. M. Mikolajczuk, P. Balczewski and S. Grzejszcak, *Synthesis* 127 (1980).
10. R. V. Honeychock, M. O. Okoroafor, L.-H. Shen and C. H. Brubaker, *Organometallics* **5**, 482 (1986).
11. See Section 3.3.1, Ref. (*15*).

12.3. REACTION WITH SULPHUR HALIDES AND RELATED COMPOUNDS

In principle, the reaction between an organolithium compound and sulphur dichloride could give a sulphenyl chloride or a thioether:

$$SCl_2 \xrightarrow{RLi} RSCl \xrightarrow{RLi} RSR$$

In practice, it is not possible to stop the reaction cleanly at the first stage, though if a sulphenyl halide is available its reaction with an organolithium compound efficiently converts it into an unsymmetrical thioether. The reaction of a dilithium compound with sulphur dichloride is useful for the synthesis of sulphur heterocycles.

As an alternative to sulphenyl halides for the synthesis of thioethers, the somewhat more readily available thiocyanates have been tried, with some success:

$$R^1Li + R^2SCN \longrightarrow R^1SR^2 + LiCN$$

It might be supposed that a reaction of two equivalents of an organolithium compound with disulphur dichloride might give a dialkyl disulphide, but the subsequent cleavage of the disulphide (Section 12.2) is too rapid (*2*).

TABLE 12.3
Reactions of Organolithium Compounds with Sulphur Dichloride, Phenylsulphenyl Chloride and Thiocyanates

Sulphur compound	Organolithium compound	Product (yield %)	Ref.
SCl_2	$Bu^tC{\equiv}CLi$	$(Bu^tC{\equiv}C)_2S$ (82–86)	(1)
SCl_2	C_6F_5Li	$(C_6F_5)_2S$ (60)	(2)
SCl_2	[2-lithio-N-(lithio-t-butylmethylidene)aniline]	2-Bu^t-benzothiazole (65)	(3)
SCl_2	bis(3,5-dichloro-4-lithiopyridyl)	tetrachloro-dipyrido-thiophene S (54)	(4)
$PhSCl^a$	$RCH(SiMe_3)Li$	$RCH(SiMe_3)SPh$ (51, R = $PhCH_2$; 53, R = Me)	(5)
MeSCN	$H_2C{=}CHCH{=}C(Li)(SMe)$	$H_2C{=}CHCH{=}C(SMe)_2$ (88)	(6)
Bu^nSCN	PhLi	$PhSBu^n$ (90)	(7)

^{a} For reactions of phenylselenenyl halides with lithium enolates see Ref. (8).

Some examples of syntheses involving reactions of organolithium compounds with sulphur dichloride, a sulphenyl chloride and thiocyanates, are shown in Table 12.3.

References

1. L. Brandsma and H. D. Verkruijsse, *Synthesis of Acetylenes, Allenes and Cumulenes*, p. 60, Elsevier, Amsterdam, 1981.
2. R. D. Chambers, J. A. Cunningham and D. A. Pyke, *Tetrahedron* **24**, 2783 (1968).
3. H. M. Walborsky and P. Ronman, *J. Org. Chem.* **43**, 731 (1978).

4. See Section 3.1.3, Ref. (*34*).
5. D. J. Ager, *J. Chem. Soc. Perkin Trans. 1* 1131 (1983).
6. R. H. Everhardus, R. Gräfing and L. Brandsma, *Recl. Trav. Chim. Pays-Bas* **97**, 69 (1978).
7. S. Gronowitz and R. Håkansson, *Arkiv Kemi* **17**, 73 (1961).
8. H. J. Reich, J. M. Renga and I. L. Reich, *J. Am. Chem. Soc.* **97**, 5434 (1975).

12.4. REACTION WITH COMPOUNDS CONTAINING SULPHUR–OXYGEN DOUBLE BONDS

Reactions of organolithium compounds with sulphonyl halides,* sulphuryl chloride, sulphinyl halides and thionyl chloride are much less satisfactory than might be supposed as routes to sulphones and sulphoxides, although exceptions to this generalization have been reported (see General Refs A and D(ii)). Alkanesulphonyl chlorides may be obtained in modest yields by the reaction of alkyllithium compounds with sulphuryl chloride at low temperatures (*1*):

$$RLi + SO_2Cl_2 \longrightarrow RSO_2Cl + LiCl$$

In contrast, as noted in Section 8.2, reactions of arenesulphonates with organolithium compounds often give sulphones rather than ethers (*2*):

$$ArSO_2OR^1 + R^2Li \longrightarrow ArSO_2R^2 + LiOR^1$$

The analogous reaction with sulphinate esters, giving sulphoxides, proceeds with inversion of configuration at sulphur (*3*). Thus a reaction of the (*S*)-(−)-menthyl ester [1] gives an enantiomerically pure chiral sulphoxide (*4*):

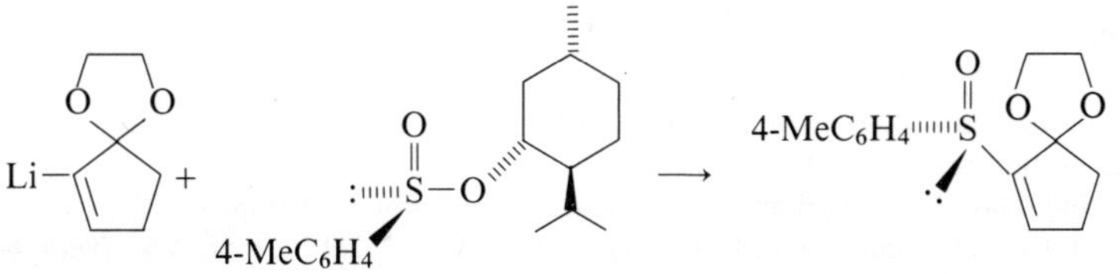

Under appropriate conditions organolithium compounds react with sulphur dioxide to give lithium sulphinates in excellent yields (*5*, *6*). Side-reactions, like those affecting the analogous reaction with carbon dioxide (Section 6.4), can be minimized by adding the organolithium compound to

---

* See Chapter 13 for reactions of sulphonyl halides leading to organic halides.

an excess of sulphur dioxide at a low temperature, as in the following example (5). The lithium sulphinate may be used for various further reactions, either *in situ* or following its isolation:

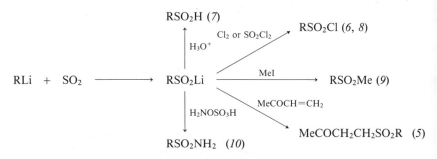

Reaction with sulphur dioxide; lithium n-butylsulphinate (5)

$$Bu^nLi + SO_2 \longrightarrow Bu^nSO_2Li$$

Sulphur dioxide (10 ml, 230 mmol) is condensed in a dry flask at $-78°$ and cold ether (40 ml) is added. n-Butyllithium (*ca*. 2.5 M in hexane, 24.5 mmol) is added dropwise during 10 min. The mixture is stirred at $-78°$ for a further 30 min and then left at room temperature for 24 h. Evaporation in a rotary evaporator leaves lithium n-butylsulphinate as a white powder (*ca*. 3.2 g).[a]

[a] The recorded yield was 3.23 g (102%), and presumably still contained traces of solvent. A subsequent reaction with butenone gave the product of Michael addition in 85% yield.

References

1. H. Quast and F. Kees, *Synthesis* 489 (1974).
2. See Section 8.2, Ref. (*9*).
3. L. Colombo, G. Gennari and E. Narisano, *Tetrahedron Lett.* 3861 (1978).
4. M. Hulce, J. P. Mallamo, L. L. Frye, T. P. Kogan and G. H. Posner, *Org. Synth.* **64**, 196 (1986).
5. H. W. Pinnick and M. A. Reynolds, *J. Org. Chem.* **44**, 160 (1979).
6. T. Hamada and O. Yonemitsu, *Synthesis* 852 (1986).
7. S. Gronowitz, *Arkiv Kemi* **13**, 269 (1958).
8. A. G. Shipov and Yu. I. Baukov, *J. Gen. Chem. USSR (Engl. Transl.)* **49**, 1112 (1979).
9. S. Asada and N. Tokura, *Bull. Chem. Soc. Jpn* **43**, 1256 (1970).
10. S. L. Graham and T. H. Scholz, *Synthesis* 1031 (1986).

—13—
Formation of Carbon–Halogen Bonds via Organolithium Compounds

In general, the methods for converting organolithium compounds into the corresponding halides may be regarded as involving reactions with sources of positive halogen—either elemental halogen or a compound in which displacement of an anionic group is favoured:

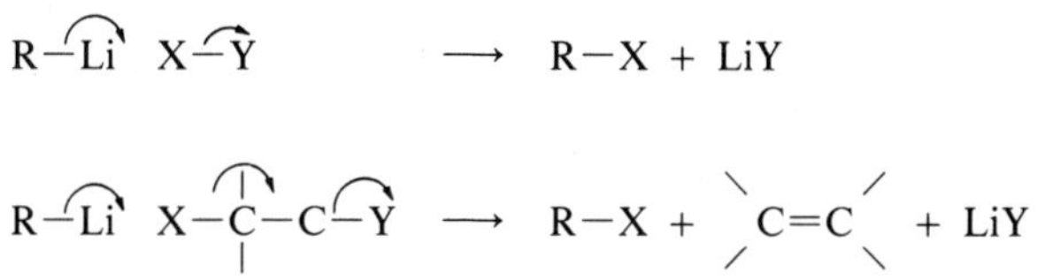

$$R\text{—Li} \quad X\text{—Y} \longrightarrow R\text{—}X + LiY$$

$$R\text{—Li} \quad X\text{—}\underset{|}{C}\text{—}C\text{—}Y \longrightarrow R\text{—}X + \overset{\diagdown}{\underset{\diagup}{C}}=\overset{\diagup}{\underset{\diagdown}{C}} + LiY$$

In the case of fluorine neither of these types of reaction is readily available. Of the reagents that have been tried, perchloryl fluoride ($FClO_3$) seemed promising (see General Ref. A), but was prone to side-reactions and was hazardous (*1*). Alternatives that have been advocated as more convenient and safer include dinitrogen difluoride (*2*), *N*-fluoro-*N*-t-butyl-*p*-toluenesulphonamide (*3*) and *N*-fluoroquinuclidinium fluoride (*4*), but none has yet been thoroughly evaluated.

The reaction of organolithium compounds with chlorine may also present an explosion hazard (*5*), but several reasonably satisfactory alternatives are available, though the yields are sometimes only modest. Some examples are listed in Table 13.1. Probably the best of them is hexachloroethane, and a detailed example of the use of this reagent follows.

For the formation of carbon–bromine bonds, reaction with the element is usually satisfactory except when other functional groups susceptible to bromination are present, in which case reagents such as those listed in Table 13.1 are available.

For iodination, reaction with the element is again usually satisfactory (as in the example below), but coupling of the organic group is sometimes observed—and may indeed be a useful reaction in its own right (*16*).

TABLE 13.1
Formation of Carbon–Chlorine, Carbon–Bromine and Carbon–Iodine Bonds via Organolithium Compounds

Halogen reagent	Organolithium compound	Product (yield %)	Ref.
Carbon–Chlorine Bonds			
C_2Cl_6	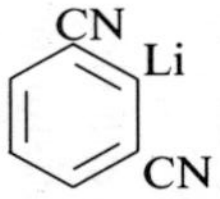 2-Li-1,3-(CN)$_2$-benzene	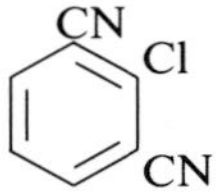 2-Cl-1,3-(CN)$_2$-benzene (77)	(6)
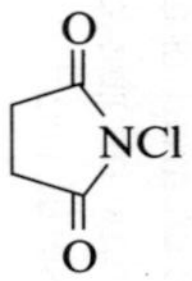 N-chlorosuccinimide	$Bu^tC{\equiv}CLi$	$Bu^tC{\equiv}CCl$ (64)	(7)
 4-MeC$_6$H$_4$SO$_2$Cl	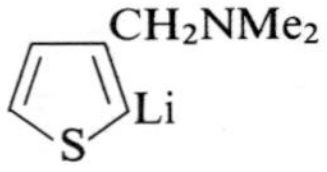 2-Li-5-(CH$_2$NMe$_2$)-thiophene	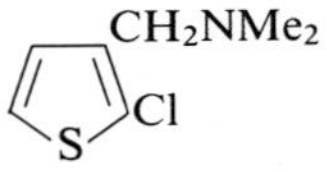 2-Cl-5-(CH$_2$NMe$_2$)-thiophene (25)	(8)
Carbon–Bromine Bonds			
Br_2	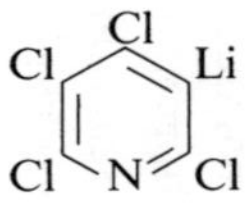 3-Li-4-Cl-2-Cl-pyridine	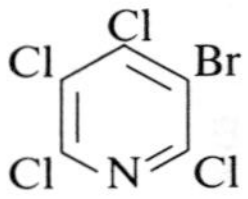 3-Br-4-Cl-2-Cl-pyridine (58)	(9)
BrCH$_2$CH$_2$Br	PhLi	PhBr (92)	(10)

TABLE 13.1—continued

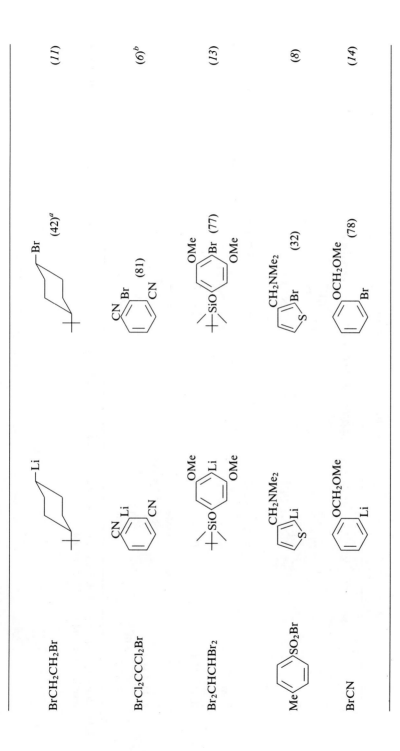

TABLE 13.1—continued

Carbon–Iodine Bonds

I_2	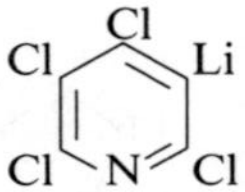 3,4-dichloro-2-chloro-5-Li-pyridine → 3,4-dichloro-2-chloro-5-I-pyridine	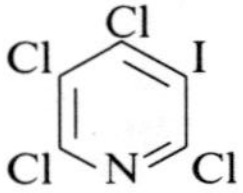 (60)	(9)
I_2	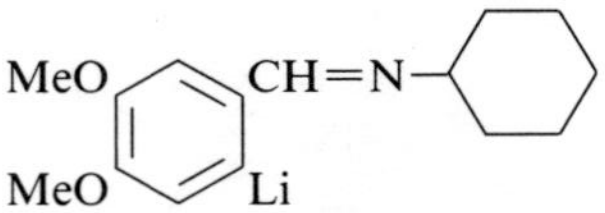 3,4-dimethoxybenzylidene-cyclohexylamine (Li → I)	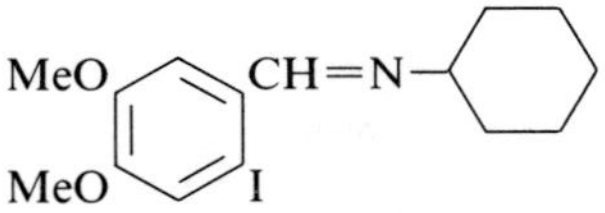 (79–85)	(14)
ICH$_2$CH$_2$I	PhLi → PhI	(83)	(10)
ICH$_2$CH$_2$Cl	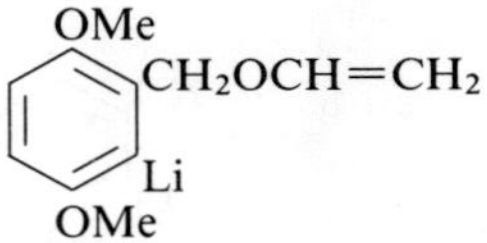 2,6-dimethoxy-3-(allyloxymethyl)phenyl Li → I	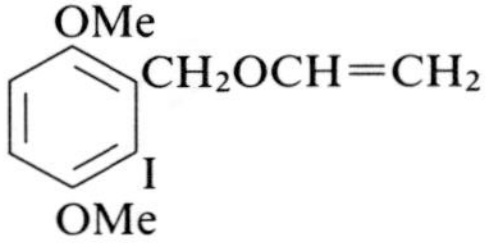 (51)	(15)

[a] Ratio *cis*:*trans* was 97:3, i.e. configuration was retained. With bromine (and similarly with chlorine and iodine) a mixture was obtained, but inversion predominated. [b] See also Ref. (*12*).

Reaction with hexachloroethane; 3-chlorofuran[a] (17)

[Structures: 3-bromofuran →(EtLi)→ 2-lithio-3-bromofuran(?) →(C_2Cl_6)→ 3-chlorofuran]

3-Bromofuran (44.1 g, 0.3 mol) is dissolved in ether (200 ml) under an atmosphere of nitrogen. The mixture is cooled to $-70°$ and stirred as ethyllithium (ca. M in ether, 0.36 mol) is added dropwise. Stirring is continued for 30 min. A solution of hexachloroethane (71.1 g, 0.3 mol) in ether (350 ml) is added during ca. 10 min, while the temperature of the mixture is kept below $-55°$. The mixture is stirred at $-70°$ for 3 h and then as it warms to room temperature. Ice-water is added to the mixture, which is then neutralized with 5 M hydrochloric acid. The phases are separated and the aqueous phase is extracted twice with ether. The combined organic phases are washed with aqueous sodium hydrogencarbonate and with water, and dried ($MgSO_4$). The ether is distilled through a column, and then a fraction, b.p. 86–87°, is collected as 3-chlorofuran (16.5 g, 54%), n_D^{22} 1.4632.

[a] 2-Chlorofuran, 2-chloroselenophene and 3-chloroselenophene were prepared similarly (17).

Reaction with elemental iodine; pentachloroiodobenzene (18)

$$C_6Cl_5Li \xrightarrow{I_2} C_6Cl_5I$$

A solution of pentachlorophenyllithium, prepared from hexachlorobenzene (22.6 g, 80 mmol) in ether (600 ml) as described on p. 29, is cooled to $-75°$ and stirred as iodine (21.0 g, 83 mmol) is added in one portion. Stirring is continued as the resulting mixture warms to room temperature. The mixture is poured into an excess of aqueous sodium thiosulphate. The organic layer is separated, dried ($MgSO_4$), and evaporated. Chromatography of the brown residue (silica, light petroleum) gives pentachloroiodobenzene (24.0 g, 80%), m.p. 209–210°.

REFERENCES

1. J. H. J. Peet and B. W. Rockett, *J. Organomet. Chem.* **82**, C57 (1974); W. Adcock and T. C. Khor, *Ibid.* **91**, C20 (1975).
2. J. Bensoam and F. Mathey, *Tetrahedron Lett.* 2797 (1977).
3. W. E. Barnette, *J. Am. Chem. Soc.* **106**, 452 (1984); S. H. Lee and J. Schwartz, *Ibid.* **108**, 2445 (1986).

4. R. E. Banks, R. A. DuBoisson and E. Tsiliopoulos, *J. Fluorine Chem.* **32**, 461 (1986).
5. J. Kattenberg, E. R. de Waard and H. O. Huisman, *Tetrahedron* **29**, 4149 (1973).
6. T. D. Krizan and J. C. Martin, *J. Org. Chem.* **47**, 2681 (1982).
7. W. Verboom, H. Westmijze, L. J. de Noten, P. Vermeer and H. J. T. Bos, *Synthesis* 296 (1979).
8. D. W. Slocum and P. L. Gierer, *J. Org. Chem.* **41**, 3668 (1976).
9. A. G. Mack, H. Suschitzky and B. J. Wakefield, *J. Chem. Soc. Perkin Trans. 1* 1472 (1979).
10. G. Wittig and G. Harborth, *Chem. Ber.* **77**, 306 (1944).
11. W. H. Glaze, C. M. Selman, A. L. Ball and L. E. Bray, *J. Org. Chem.* **34**, 641 (1969).
12. B. Abarca, R. Ballasteros, G. Jones and F. Mojarrad, *Tetrahedron Lett.* **27**, 3543 (1986).
13. B. M. Trost and M. G. Saulnier, *Tetrahedron Lett.* **26**, 123 (1985).
14. F. E. Ziegler, K. W. Fowler, W. B. Rodgers and R. T. Wester, *Org. Synth.* **65**, 108 (1987).
15. R. C. Ronald, J. M. Lansinger, T. L. Lillie and C. J. Wheeler, *J. Org. Chem.* **47**, 2541 (1982); see also Section 3.2, Ref. (*11*).
16. J. L. Belletire, E. G. Spletzer and A. R. Pinkas, *Tetrahedron Lett.* **25**, 5969 (1984).
17. S Gronowitz, A.-B. Hörnfeldt and K. Petterson, *Synth. Commun.* **3**, 213 (1973).
18. A. G. Mack, Ph.D. Thesis, Salford University, 1978.

—14—

Synthesis of Organoboron, Organosilicon and Organophosphorus Compounds from Organolithium Compounds

The reactions somewhat arbitrarily grouped here are general, usually straightforward, and useful:

$$n\text{RLi} + \text{MX}_m \longrightarrow \text{R}_n\text{MX}_{m-n} + n\text{LiX}$$

$$n\text{R}^1\text{Li} + \text{R}_m^2\text{MX}_p \longrightarrow \text{R}_n^1\text{R}_m^2\text{MX}_{p-n} + n\text{LiX}$$

$$\text{M} = \text{B, Si, P; X} = \text{halogen, OR}^3 \text{ etc.}$$

The reactions with boron derivatives are commonly used, for example, to achieve indirect conversion of the organolithium compound into the corresponding hydroxy-compound, the reactions with silicon derivatives to characterize organolithium compounds, and the reactions with phosphorus derivatives to prepare phosphine ligands.

Although the reactions are generally straightforward, clean stepwise alkylation of polyhalides etc. is sometimes a problem, but is usually achievable by careful attention to the stoichiometry, mode of addition and temperature of reaction. The use of substrates bearing mixed ligands may be advantageous.

14.1. SYNTHESIS AND OXIDATION OF ORGANOBORON COMPOUNDS

Some examples of reactions of organolithium compounds with boron derivatives are listed in Table 14.1 and others are tabulated in General Ref. A. In the case of boron, the reactions represented above have to be extended. Use of an excess of the organolithium compound, or reaction of an organolithium compound with a trialkylborane, gives a lithium tetraalkylborate:

$$4\text{RLi} + \text{BX}_3 \longrightarrow \text{LiBR}_4 \longleftarrow \text{R}_3\text{B} + \text{RLi}$$

TABLE 14.1

Synthesis of Organoboron Compounds from Organolithium Compounds

Organolithium compound	Boron reagent	Product (yield %)	Ref.
Bu^nLi	$B(OBu^n)_3$	$Bu^nB(OBu^n)_2$ (60)	(2)
$2Bu^tLi$	$B(SMe)_3$	Bu^t_2BSMe (70)	(3)
$2Bu^tLi$	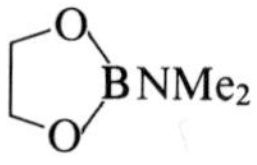 ⟨OCH₂CH₂O⟩BNMe₂	$Bu^t_2BNMe_2$ (69)	(3)
$3Bu^tLi$	$B(OMe)_3$	Bu^t_3B (56)	(4)
$PhC{\equiv}CLi$	$Ph_2BBr \cdot C_5H_5N$	$PhC{\equiv}CBPh_2 \cdot C_5H_5N$ (70)	(5)
$3PhLi$	$BF_3 \cdot OEt_2$	Ph_3B (65–70)	(6)
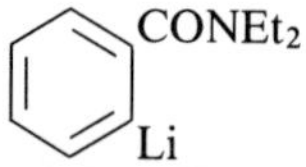 2-Li-C₆H₄-CONEt₂	$B(OMe)_3$	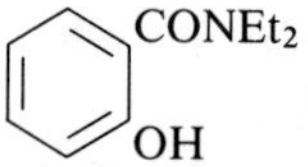 2-HO-C₆H₄-CONEt₂ (a) (56)	(7)
$3C_6F_5Li$	BCl_3	$(C_6F_5)_3B$ (30–50)	(8)
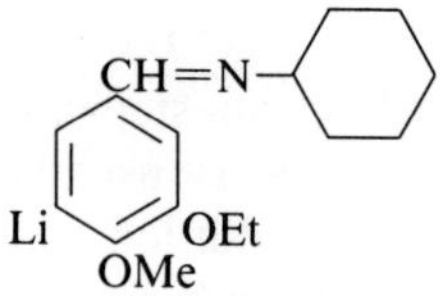 [2-Li-3-OMe-4-OEt-C₆H₂-CH=N-C₆H₁₁]	$B(OBu^n)_3$	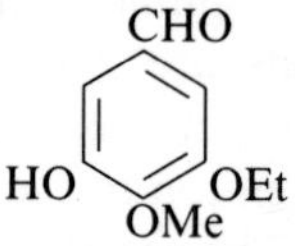 [2-HO-3-OMe-4-OEt-C₆H₂-CHO] (b) (77)	(9)
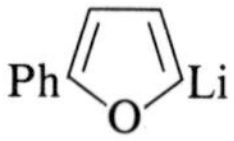 5-Ph-2-Li-furan	$ClB(OMe)_2$	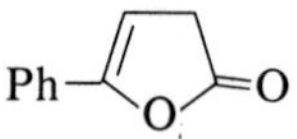 5-Ph-furan-2(5H)-one (c) (91)	(10)
2 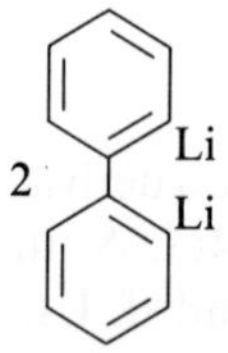 × (2,2'-dilithiobiphenyl)	$BF_3 \cdot OEt_2$	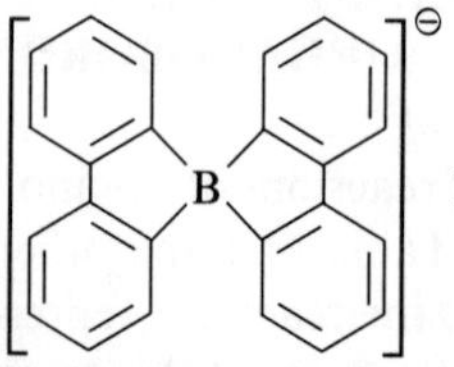 spirobi(dibenzoborole) anion Li⊕ (45–50)	(11)

[a] From oxidation of organoboron product with hydrogen peroxide. [b] From oxidation and hydrolysis of organoboron product. [c] From oxidation of organoboron product with m-chloroperoxybenzoic acid.

14.1. SYNTHESIS AND OXIDATION OF ORGANOBORON COMPOUNDS

The most commonly employed boron reagents are boron halides and borate esters, but, as is apparent from Table 14.1, a variety of other substrates may be used.

Table 14.1 includes two examples where the organoboron product was oxidized, and another example is described in detail below. For such reactions hydrogen peroxide has been routinely used as the oxidant. However, it may be advantageous to use buffered m-chloroperoxybenzoic acid, as in the synthesis of 5-phenyl-2($3H$)-butenolide shown in Table 14.1. A potentially useful one-step equivalent of these procedures has been reported, in which the reagent is 2-t-butylperoxy-1,3,2-dioxaborolane [1] (*1*):

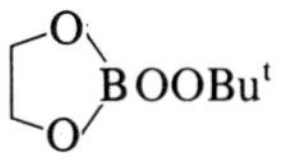

[1]

Synthesis of benzo[b]thiophen-2(3H)-one by formation and oxidation of an organoboron compound (12)

In this example the hydroxy-compound formed by oxidation of the organoboron intermediate tautomerizes to the corresponding carbonyl compound:

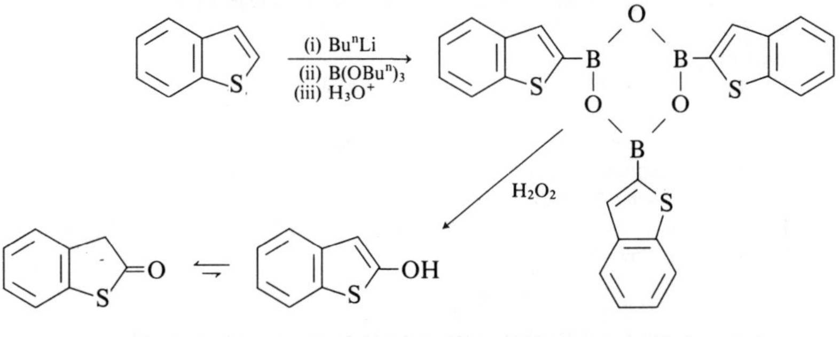

Benzo[b]thiophene (40.2 g, 0.3 mol) is dissolved in ether (450 ml) in a 2 l three-necked flask fitted with a septum, a pressure-equalizing dropping funnel and a stirrer, and furnished with an inert atmosphere. The solution is cooled in ice and stirred as n-butyllithium (*ca.* 2 M in hexane, 0.3 mol) is added dropwise. The mixture is stirred at room temperature for 1 h and then re-cooled in ice. Stirring and cooling are continued as a solution of tri-n-butyl borate (82.8 g, 0.36 mol) in ether (100 ml) is added dropwise during 10 min. The resulting mixture is stirred at room temperature for 1 h: 2 M hydrochloric acid is then added. The organic layer is separated and the aqueous layer is extracted with ether. The combined organic layers are extracted thoroughly with 2 M sodium hydroxide. The extract is acidified with concentrated hydrochloric acid and extracted with ether. The ethereal extract is washed with water, dried (MgSO$_4$) and evaporated. The solid residue is dried at 140° to leave 1,3,5-tris(benzo[b]thien-2-yl)boroxine (45.5 g, 95%)^a.

The boroxine (18.7 g, 39 mmol) is added in small portions to stirred 30% w/v hydrogen peroxide (20 ml) at such a rate that the temperature of the reaction mixture does not exceed 50°. When all the boroxine has been added, the mixture is stirred at 70° for 15 min. The mixture is cooled, diluted with water and extracted thoroughly with chloroform. The combined extracts are washed with water, dried, and evaporated, leaving benzo[b]thiophen-2(3H)-one (13.3 g, 76%), m.p. 43-44° (from light petroleum, b.p. 40-60°).

^{a} M.p. ca. 250° (from benzene).

References

1. See Section 11.1, Ref. (6).
2. P. B. Brindley, W. Gerrard and M. F. Lappert, *J. Chem. Soc.* 2956 (1955).
3. H. Nöth, H. Prigge and T. Taeger, *Organomet. Synth.* **3**, 441 (1986).
4. H. Nöth, H. Prigge and T. Taeger, *Organomet. Synth.* **3**. 459 (1986).
5. J. J. Eisch, *Organomet. Synth.* **2**, 130 (1981); *cf.* J. Soulié and P. Cadiot, *Bull. Soc. Chim. Fr.* 1981 (1966).
6. J. J. Eisch, *Organomet. Synth.* **2**, 128 (1981); *cf.* G. Wittig and P. Raff, *Justus Liebigs Ann. Chem.* **573**, 195 (1951).
7. See Section 3.2, Ref. (37).
8. A. G. Massey and A. J. Park, *Organomet. Synth.* **3**, 461 (1986).
9. P. Jacob and A. T. Shulgin, *Synth. Commun.* **11**, 969 (1981).
10. A. Pelter and M. Rowlands, *Tetrahedron Lett.* **28**, 1203 (1987).
11. J. J. Eisch, *Organomet. Synth.* **2**, 134 (1981); *cf.* J. J. Eisch and R. J. Wilcsek, *J. Organomet. Chem.* **71**, C21 (1974).
12. R. P. Dickinson and B. Iddon, *J. Chem Soc. (C)* 1926 (1970); R. P. Dickinson, Ph.D. Thesis, Salford, 1969.

14.2. SYNTHESIS OF ORGANOSILICON COMPOUNDS

The majority of syntheses of organosilicon compounds from organolithium compounds have employed silicon halides, especially trialkylsilyl chlorides, as substrates. Chlorotrimethylsilane is routinely used as a convenient reagent for characterizing organolithium compounds as well as for preparing desired trimethylsilyl compounds; the following procedure is typical. Further examples are listed in Table 14.2. These and other reactions with halosilanes are reviewed in General Ref. A (and see Ref. (1)).

The use of a dihalide for preparing a silicon heterocycle is illustrated by

14.2. SYNTHESIS OF ORGANOSILICON COMPOUNDS

TABLE 14.2
Reactions of Organolithium Compounds with Chlorotrimethylsilane

Organolithium compound	Product (yield %)	Ref.
MeOCH$_2$Li	MeOCH$_2$SiMe$_3$ (80)	(3)
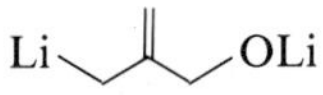 Li⌒⨀⌒OLi	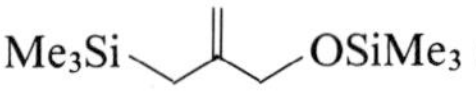 Me$_3$Si⌒⨀⌒OSiMe$_3$ (52)	(4)
PhSCH$_2$Li	PhSCH$_2$SiMe$_3$ (95)	(5)
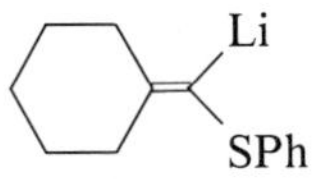 cyclohexylidene(Li)(SPh)	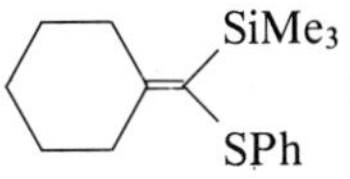 cyclohexylidene(SiMe$_3$)(SPh) (96)	(6)
BuC≡CLi	BuC≡CSiMe$_3$ (71)	(7)
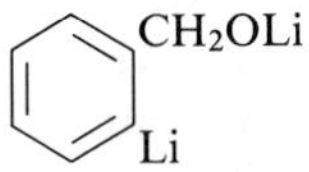 2-Li-C$_6$H$_4$-CH$_2$OLi	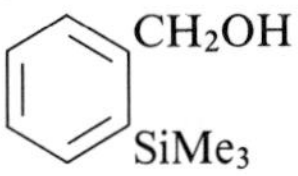 2-SiMe$_3$-C$_6$H$_4$-CH$_2$OH (63)	(8)
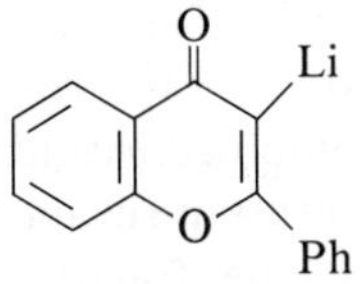 3-Li-2-Ph-chromone	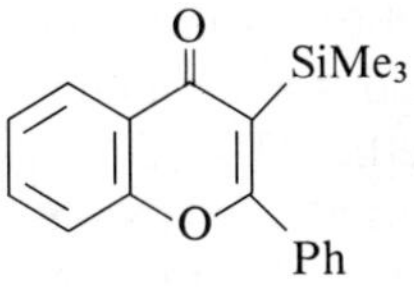 3-SiMe$_3$-2-Ph-chromone (86)	(9)
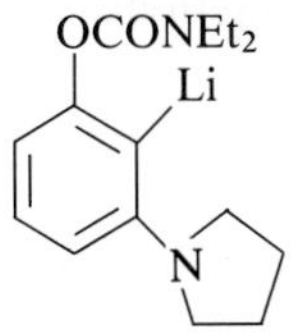 2-OCONEt$_2$-3-pyrrolidinyl-C$_6$H$_3$-Li	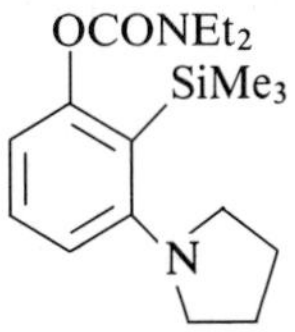 2-OCONEt$_2$-3-pyrrolidinyl-C$_6$H$_3$-SiMe$_3$ (96)	(10)

the following reaction (2). An analogous reaction with silicon tetrachloride gave the spirosilane in 22% yield.

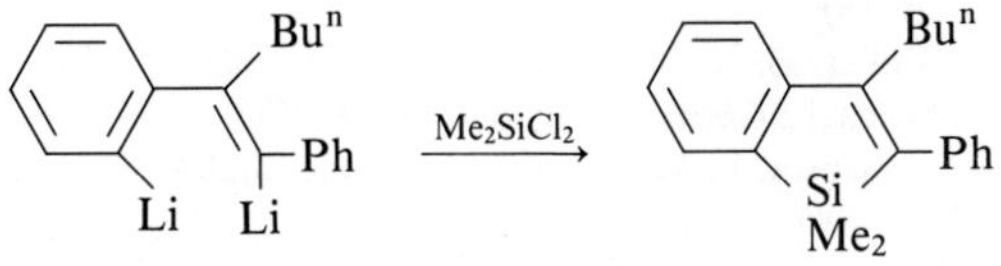

2-(α-Li-α-Ph-CH=)C$_6$H$_4$-Li with Bu^n substituent → (Me$_2$SiCl$_2$) → benzosilole with Bu^n, Ph, SiMe$_2$ ring

Other groups besides halogen (including hydrogen) may be displaced from

silicon by organolithium compounds, but these reactions are of minor value in synthesis (see General Ref. A and Ref. (1)).
The stereochemistry of nucleophilic substitution at silicon differs from that at carbon. For reactions with organolithium compounds, either inversion or *retention* of configuration is usually observed, depending on the nature of the organolithium compound and of the leaving group. More detailed discussion of such reactions will be found in reviews (1, 11).

Reaction with chlorotrimethylsilane; 1-phenylthio-1-trimethylsilylethene (5)

$$CH_2=CHSPh \xrightarrow{Bu^nLi} CH_2=C\genfrac{}{}{0pt}{}{SPh}{Li} \xrightarrow{Me_3SiCl} CH_2=C\genfrac{}{}{0pt}{}{SPh}{SiMe_3}$$

A mixture of THF (150 ml), n-butyllithium (*ca.* 1.5 M in hexane, 0.1 mol) and TMEDA (15.1 ml, 11.6 g, 0.1 mol) is cooled to −90°. The mixture is stirred and its temperature is kept below −80° as phenylthioethene (13.6 g, 0.1 mol) is added slowly. When the addition is complete, stirring is continued for 1 h at −90°. Chlorotrimethylsilane (15.2 ml, 13.0 g, 0.12 mol) is added and the reaction mixture is allowed to warm to room temperature. The mixture is poured into saturated aqueous ammonium chloride (300 ml). The resulting mixture is extracted with ether (3 × 150 ml) and the combined extracts are washed with water (100 ml) and dried (Na_2SO_4). The solvents are evaporated under reduced pressure and the residue is distilled, the fraction b.p. 130–134°/15 mmHg being collected as 1-phenylthio-1-trimethylsilylethene (17.1 g, 82%).

References

1. D. A. Armitage, *Comprehensive Organometallic Chemistry*, ed. G. Wilkinson, Vol. 2, Chap. 9.1, Pergamon, Oxford, 1982.
2. M. D. Rausch and L. P. Klemann, *Organomet. Synth.* **3**, 507 (1986).
3. R. F. Cunico and H. S. Gill, *Organometallics* **1**, 1 (1982).
4. See Section 3.2, Ref. (*44*).
5. D. J. Ager, *J. Chem. Soc. Perkin Trans. 1* 1131 (1983).
6. See Section 3.4, Ref. (*6*).
7. See Section 3.1.3, Ref. (*15*).
8. N. Meyer and D. Seebach, *Chem. Ber.* **113**, 1304 (1980).
9. A. M. B. S. R. C. S. Costa, F. M. Dean, M. A. Jones and R. S. Varma, *J. Chem. Soc. Perkin Trans. 1* 799 (1985).
10. M. Skrowońska-Ptasinska, W. Verboom and D. N. Reinhoudt, *J. Org. Chem.* **50**, 2690 (1985).
11. R. G. P. Corriu and C. Guerin, *Adv. Organomet. Chem.* **20**, 265 (1982).

14.3. SYNTHESIS OF ORGANOPHOSPHORUS COMPOUNDS

14.3.1. Phosphorus(III) Compounds

Reactions of organolithium compounds with phosphorus trihalides, alkyldihalophosphines and dialkylhalophosphines are so straightforward as to require little comment. Besides the following example and those listed in Table 14.3, many examples are tabulated in General Ref. A.

Table 14.3 also includes examples of displacement of alkoxide, which is occasionally used as an alternative to displacement of halide. This reaction with chiral O-menthyl phosphinites proceeds with inversion of configuration and has been used to prepare chiral phosphines (9):

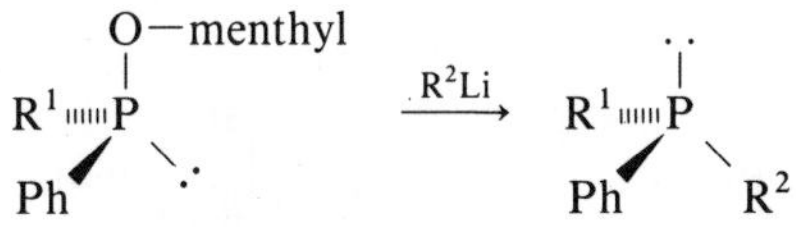

$$R^1\text{(}\cdots\text{)}P(\text{O-menthyl})(Ph) \xrightarrow{R^2Li} R^1\text{(}\cdots\text{)}P(Ph)(R^2)$$

R^1 = Me, Et; R^2 = Me, Pr^n, Bu^n, Bu^t

Reaction with a halophosphine;
[2-(methylthio)phenyl] diphenylphosphine (5)

$$\text{(o-Br-C}_6\text{H}_4\text{-SMe)} \xrightarrow{BuLi} \text{(o-Li-C}_6\text{H}_4\text{-SMe)} \xrightarrow{Ph_2PCl} \text{(o-PPh}_2\text{-C}_6\text{H}_4\text{-SMe)}$$

1-Bromo-2-methylthiobenzene (o-bromothioanisole) (38.5 g, 0.19 mol) is placed in a 1 l three-necked flask fitted with a condenser, a pressure-equalizing dropping funnel and a mechanical stirrer, and furnished with an atmosphere of nitrogen. Ether (200 ml) is added, and the solution is purged with nitrogen for 30 min. The solution is stirred vigorously at 0° as n-butyllithium (ca. 1.5 M in hexane, 0.19 mol) is added dropwise during 2 h. The mixture is stirred for a further 1 h at 0°. A solution of vacuum-distilled chlorodiphenylphosphine (diphenylphosphinous chloride) (42 g, 0.19 mol) in ether (125 ml) is added dropwise to the stirred solution during 3 h. A white precipitate forms. 0.2 M hydrochloric acid (125 ml) is added and the mixture is stirred to ensure that all the inorganic salts dissolve. The white crystalline product is collected by filtration, washed with water, ethanol and ether, and dried in a desiccator. The yield is 43 g (74%), m.p. 101–102°.

TABLE 14.3

Synthesis of Organophosphorus Compounds from Organolithium Compounds

Organolithium compound	Phosphorus reagent	Product (yield %)	Ref.
Phosphorus(III)			
MeSCH$_2$Li	P(OPh)$_3$	(MeSCH$_2$)$_3$P (54)	(1)
Ph$_2$PCH$_2$Li	Ph$_2$POPh	(Ph$_2$P)$_2$CH$_2$ (ca. 50)	(2)
PhCH(Li)SMe	Ph$_2$PCl	PhCH(SMe)(PPh$_2$) (45)	(3)
2-(SPr^i)C$_6$H$_4$Li	Ph$_2$PCl	2-(SPr^i)-1-(PPh$_2$)C$_6$H$_4$ (66)	(4)
2-(SMe)C$_6$H$_4$Li	PCl$_3$	[2-(SMe)C$_6$H$_4$]$_3$P (89)	(5)
 2-(SMe)C$_6$H$_4$Li	PhPCl$_2$	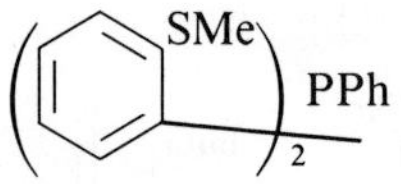 [2-(SMe)C$_6$H$_4$]$_2$PPh (89)	(5)
Phosphorus(V)			
2,4,6-Bu$^t_3$C$_6$H$_2$Li	POCl$_3$	(2,4,6-Bu$^t_3$C$_6$H$_2$)$_2$P(O)Cl (49)	(6)
1-CH$_2$Ph-2-Ph-imidazol-5-yl-Li	(EtO)$_2$P(O)Cl	1-CH$_2$Ph-2-Ph-5-[P(O)(OEt)$_2$]-imidazole (54)	(7)
2-(AsPh$_2$)C$_6$H$_4$Li	Ph$_2$P(S)Cl	2-(AsPh$_2$)-1-[P(S)Ph$_2$]C$_6$H$_4$ (46)	(8)

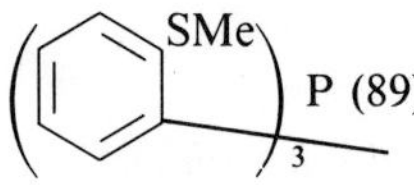

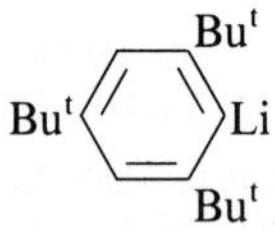

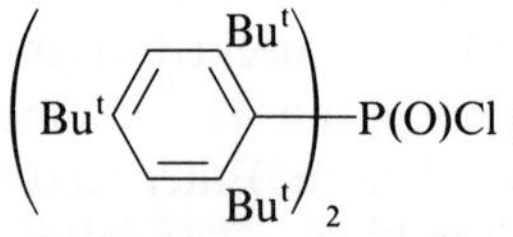

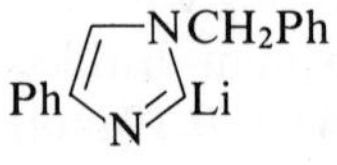

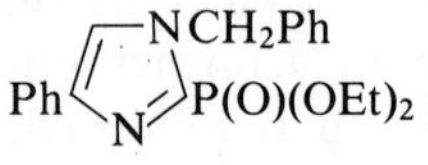

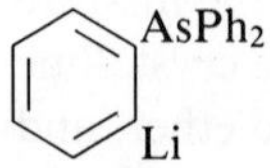

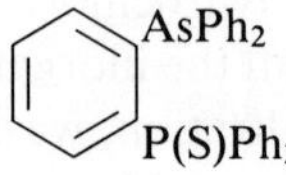

14.3.2. Phosphorus(v) Compounds

In reactions of organolithium compounds with phosphorus(v) compounds, α-deprotonation competes with substitution at phosphorus; the Horner–Wadsworth–Emmons synthesis (see Section 6.1.3, p. 75) provides an example (*10, 11*). Nevertheless, with a good leaving group such as halide and/or in the absence of α-hydrogen atoms, products of substitution may be obtained in fair yields. Examples of the types of substrate that have been used are shown below* (see General Refs A and D(ii)), and some specific examples are listed in Table 14.3:

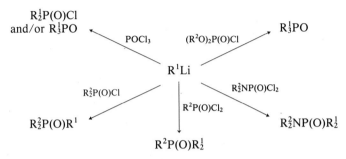

Besides the reactions shown here, involving displacement of chloride, some reactions involving displacement of alkoxide are known (*11*), but in this case substitution at carbon competes, particularly when substitution at phosphorus is sterically hindered (*12*):

The stereochemistry of substitution at phosphorus(v) has been a subject of controversy (*13*). Substitution by organolithium compounds usually proceeds with predominant inversion of configuration (*14*), though retention has been observed.

References

1. D. J. Peterson, *J. Org. Chem.* **32**, 1717 (1967).
2. D. J. Peterson, *J. Organomet. Chem.* **8**, 199 (1967).
3. See Section 3.2, Ref. (*46*).

*Similar reactions occur with many sulphur analogues.

4. See Section 3.2, Ref. (*14*).
5. See Section 3.1.3, Ref. (*27*).
6. M. Yoshifugi, I. Shima and N. Inamoto, *Tetrahedron Lett.* 3963 (1979).
7. D. K. Anderson, J. A. Sikorski, D. B. Reitz and L. T. Pilla, *J. Heterocycl. Chem.* **23**, 1257 (1986).
8. P. Nicpon and D. W. Meek, *Inorg. Chem.* **6**, 145 (1967).
9. M. Mikolajczuk, J. Omelanczuk and W. Perlikowska, *Tetrahedron* **35**, 1531 (1979); J. Omelanczuk and M. Mikolajcuk, *J. Chem. Soc. Chem. Commun.* 24, (1980).
10. See Section 6.1.3, Ref. (*4*).
11. M.-P. Teulade and P. Savignac, *Tetrahedron Lett.* **28**, 405 (1987).
12. H. Gilman and B. J. Gaj, *J. Am. Chem. Soc.* **82**, 6326 (1960).
13. *Organophosphorus Chemistry*, Specialist Periodical Reports, Royal Society of Chemistry, London, Vols 1–17 (1970–1987).
14. See e.g. R. A. Lewis and K. Mislow, *J. Am. Chem. Soc.* **91**, 7009 (1969); W. B. Farnham, R. K. Murray and K. Mislow, *Ibid.* **92**, 5809 (1970).

—15—
Organolithium Compounds in the Synthesis of Other Organometallic Compounds

The reaction of an organolithium compound with a derivative—usually the halide—of a less-electropositive metal may be applied to the synthesis of organic derivatives of most metals:

$$n\text{RLi} + \text{MX}_m \longrightarrow \text{R}_n\text{MX}_{m-n} + n\text{LiX}$$

In many cases an extension can be used to prepare "ate" complexes:

$$(m + 1)\text{RLi} + \text{MX}_m \longrightarrow \text{LiMR}_{m+1} \longleftarrow \text{RLi} + \text{R}_m\text{M}$$

There are many variations on these reaction schemes, and detailed discussion of "non-standard" reactions would be beyond the scope of this book. Most procedures are straightforward, however. The advantage of organolithium compounds lies in their high nucleophilic reactivity, which enables very mild conditions to be used to prepare unstable products. Drawbacks occasionally encountered are associated with their high basicity and/or their capacity to act as one-electron donors.

The preparations described in detail below, one of a derivative of a main-group metal and one of a derivative of a transition metal, are typical of reactions giving well-defined products. A selection of other examples is listed in Table 15.1, and others are tabulated in General Ref. A.

Although the first general reaction above refers to the synthesis of organic derivatives of less electropositive metals, it may also occasionally be applied to the synthesis of derivatives of metals more electropositive than lithium, as in the preparation of organopotassium reagents noted in Section 3.2. The preparation of these reagents is typical of a large number of applications in which a metal derivative is added to an organolithium reagent to give a new reagent with desirable properties. The new reagent is commonly used *in situ* and its constitution is often unknown. Some leading references to such applications are given in Table 15.2. Examples of some of these have been referred to in earlier sections, as indicated.

TABLE 15.1
Synthesis of Other Organometallic Compounds from Organolithium Compounds

Organolithium compound	Substrate	Product (yield %)	Ref.
Me$_2$NCH$_2$CH$_2$Li	ZnCl$_2$	(Me$_2$NCH$_2$CH$_2$)$_2$Zn (61)	(1)
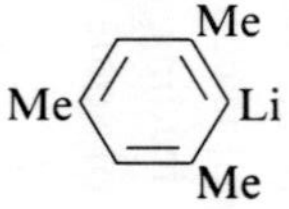 2,6-Me$_2$C$_6$H$_3$Li (Me, Li, Me)	ZnCl$_2$	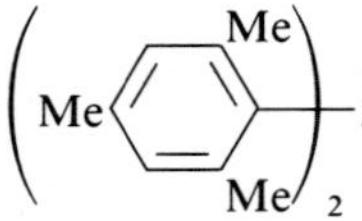 (2,6-Me$_2$C$_6$H$_3$)$_2$Zn (33)[a]	(2)
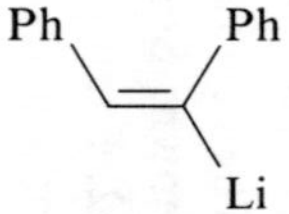 Ph(Li)C=CHPh	HgCl$_2$	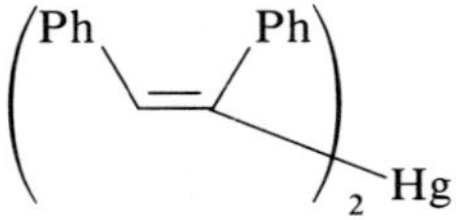 (PhC=CHPh)$_2$Hg (55)[b]	(3)
Me$_3$SiCH$_2$Li	AlBr$_3$	(Me$_3$SiCH$_2$)$_3$Al (85)	(4)
(Me$_2$P)$_2$CHLi	Me$_2$AlCl	(Me$_2$P)$_2$CHAlMe$_2$ (86)	(5)[c]
C$_6$F$_5$Li	TlCl$_3$	(C$_6$F$_5$)$_2$TlCl (83)	(6)
PhLi	Ph$_2$TlBr	Ph$_3$Tl (65–70)	(7)
PhLi	GeCl$_4$	Ph$_4$Ge (88)	(8)
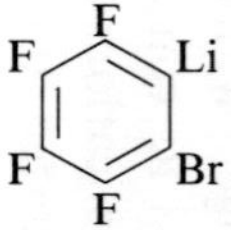 2-Br-3,4,5,6-F$_4$C$_6$Li	Ph$_2$GeCl$_2$	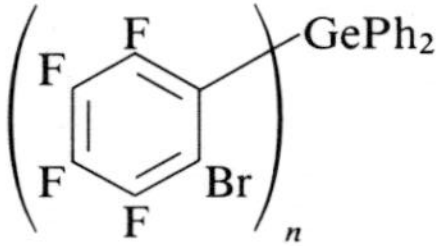 (—C$_6$F$_4$—GePh$_2$—)$_n$ from 2-Br-C$_6$F$_4$ (76)	(9)[d]

TABLE 15.1—continued

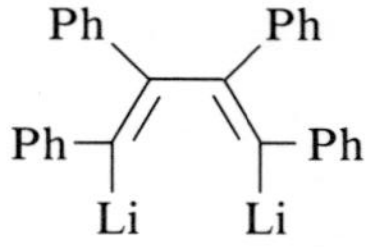 Ph−C(Ph)=C(Ph)−C(Li)(Ph)−Li with extra Ph/Li	Me$_2$SnCl$_2$	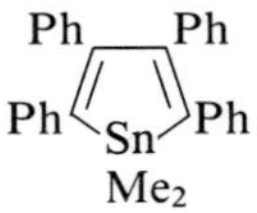 Ph−C(Ph)=C(Ph)−Sn(Me$_2$)−C(Ph)=Ph (76)[e]	(10)
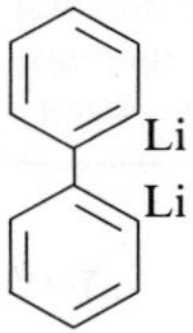 biphenyl-2,2'-diyl dilithium	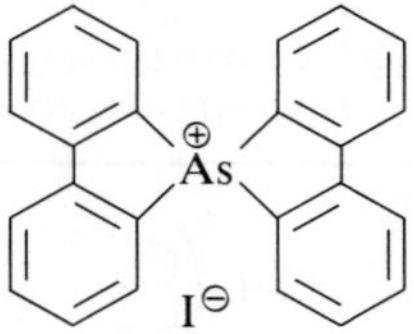 [dibenzarsole cation] I[−]	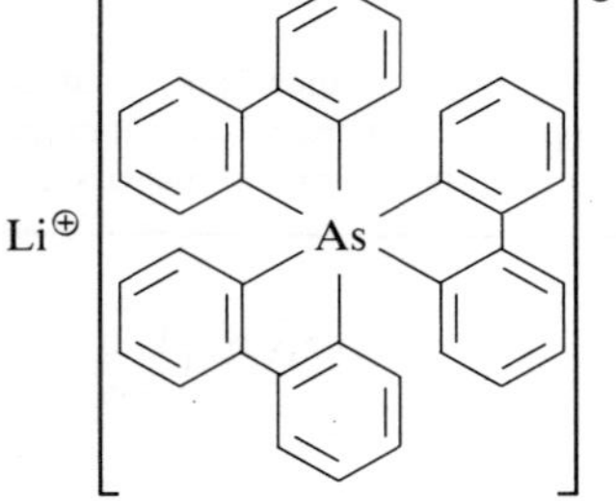 [spiro-tetra(biphenyl-2,2'-diyl)arsenate] Li[+] (75)[f]	(11)
Me$_3$SiCH$_2$Li	Cu$_2$I$_2$	(Me$_3$SiCH$_2$Cu)$_n$ (70)	(12)
[13]CH$_3$Li	IAgPBu$^n_3$	[13]CH$_3$AgPBu$^n_3$ (g)	(13)
C$_6$F$_5$Li	AgOAc	C$_6$F$_5$Ag (64)	(14)
C$_6$F$_5$Li	AuCl(SBu^n)	C$_6$F$_5$Au(SBu^n) (98)	(15)[h]
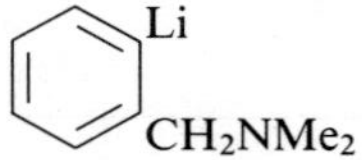 2-(Me$_2$NCH$_2$)C$_6$H$_4$Li	BrAuPPh$_3$	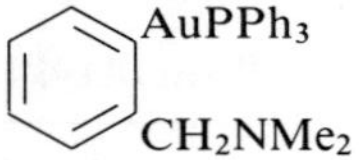 2-(Me$_2$NCH$_2$)C$_6$H$_4$AuPPh$_3$ (65)	(16)

TABLE 15.1—continued

MeLi	ClTi(OPr^i)$_3$	MeTi(OPr^i)$_3$ (95)^j	(17, 18)
Me$_3$SiCH$_2$Li	VCl$_4$	(Me$_3$SiCH$_2$)$_4$V (30)	(19)
Me$_3$CCH$_2$Li	CrCl$_3$(THF)$_3$	(Me$_3$CCH$_2$)$_4$Cr (40)^k	(20)
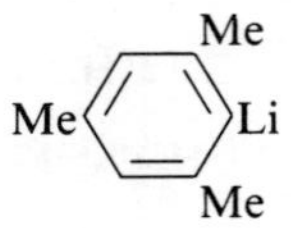 Me–(C$_6$H$_3$)(Me)–Li, Me	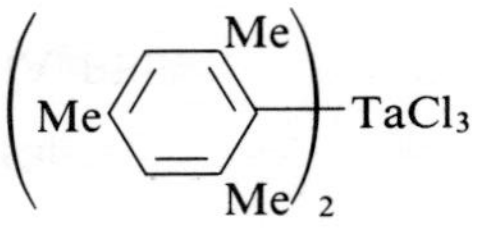 (Me)(C$_6$H$_3$)(Me)–C(Me)$_2$–TaCl$_3$	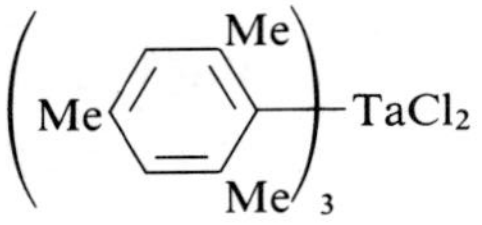 (Me)(C$_6$H$_3$)(Me)–C(Me)$_3$–TaCl$_2$ (74)	(2)
MeLi	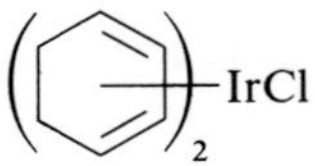 [(C$_6$H$_6$)IrCl]$_2$	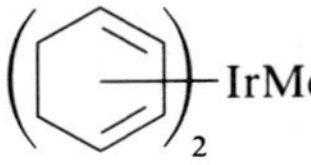 [(C$_6$H$_6$)IrMe]$_2$ (56)	(21)

^{a} Recrystallized. ^{b} The Z-isomer is prepared similarly (70%). ^{c} Other examples also described, including "ate" complexes. ^{d} Preparation of related compounds also described. ^{e} Based on diphenylacetylene, the precursor of the dilithium compound. Based on the dilithium compound, the yield is almost quantitative. ^{f} THF complex. ^{g} Thermally unstable; not isolated. ^{h} Preparation of "ate" complex [Bu$_4^n$N][Au(C$_6$F$_5$)$_2$] also described. ^{i} 69% on the smaller scale of Ref. (18). ^{k} Sublimed.

15. SYNTHESIS OF OTHER ORGANOMETALLIC COMPOUNDS

TABLE 15.2
Organometallic Reagents Prepared *in situ* from Organolithium Compounds

Metal	Ref.	Notes
Na	*(26)*	
K	*(26, 27)*	See Section 3.2
Mg	*(22)*	See text and Section 11.1
Zn	*(28, 29)*	A "Barbier" technique, assisted by ultrasound, is used
Cd	*(30)*	
Cu	*(31)*	See Sections 6.1.2 and 8.1
Ti	*(32)*	See Section 6.1.1
Hf	*(33, 34)*	Cr, Mo also mentioned
Mn	*(35)*	See Section 6.1.1
Ce	*(36–40)*	
Pd	*(41)*	See Section 8.1

The reaction with magnesium bromide, by which an organolithium compound is converted into a "Grignard reagent", is quite often used. A solution of anhydrous magnesium bromide in ether is conveniently prepared *in situ* by the reaction of magnesium with bromine *(22)*, 1,2-dibromoethane *(23)* or mercury(II) bromide *(24)*. The corresponding reactions in THF give a two-phase system, the dense lower phase containing a magnesium bromide–THF complex. *(25)*.

Methyldiphenylbismuth (42)

$$Ph_2BiCl + MeLi \longrightarrow Ph_2BiMe + LiCl$$

Chlorodiphenylbismuth (9.97 g, 25.0 mmol) is suspended in ether (100 ml). The suspension is stirred vigorously and its temperature is maintained between $-65°$ and $-70°$ as a solution of methyllithium (25.0 mmol) in ether is added during *ca.* 1 h. The resulting white suspension is then allowed to warm to room temperature, to give a colourless to pale-yellow solution and a greyish precipitate. The solvent is evaporated under reduced pressure at room temperature, and the residue is digested with light petroleum (3 × 100 ml), which is decanted each time. The combined solutions are evaporated under reduced pressure at room temperature, and the oily residue is distilled in a Kugelrohr apparatus. A small forerun (5–10% of the total) is followed by methyldiphenylbismuth (89%), b.p. 105–110°/0.02 mmHg, a colourless liquid, n_D^{20} 1.6788.

cis-*Bis[4-(trimethylsilyl)phenyl]bis(triphenylphosphine)platinum (43)*

Me$_3$Si—C$_6$H$_4$—Br $\xrightarrow{\text{Bu}^n\text{Li}}$ Me$_3$Si—C$_6$H$_4$—Li

$\xrightarrow{cis\text{-Pt(PPh}_3\text{)}_2\text{Cl}_2}$ cis-Pt(C$_6$H$_4$-4-SiMe$_3$)$_2$(PPh$_3$)$_2$

A solution of 1-bromo-4-(trimethylsilyl)benzene (4.60 g, 20.8 mmol) in ether (50 ml) is prepared in a 200 ml Schlenk flask fitted with a stirrer and a side-arm closed by a septum and furnished with an argon atmosphere. The solution is cooled to −78° and stirred vigorously as n-butyllithium (ca. 1.6 M in hexane, 20.0 mmol) is added dropwise by syringe. Stirring is continued for 2 h, by which time the solution is yellow.

A fine suspension of *cis*-dichlorobis(triphenylphosphine)platinum (2.00 g, 2.8 mmol) in ether (50 ml) is prepared by ultrasonic irradiation for 10 min, and added in one portion, in a current of argon, to the cold stirred solution of the organolithium compound. The mixture is allowed to warm to room temperature during 1 h, and stirring is continued for a further 4 h. The mixture is poured into ice–water (200 ml). The organic layer is separated and the aqueous layer is extracted with dichloromethane (3 × 200 ml). The combined organic layers are washed with water (2 × 100 ml), dried (MgSO$_4$), concentrated to one twentieth of their volume, diluted with methanol (300 ml), and cooled to −20° for 48 h. The product that crystallizes is purified by chromatography (Kieselgel, dichloromethane), giving a yield of 1.90 g (67%), m.p. 155° dec.

REFERENCES

1. K.-H. Thiele, E. Langguth and G. E. Müller, *Z. Anorg. Allg. Chem.* **462**, 152 (1980).
2. P. R. Sharp, D. Astruc and R. D. Schrock, *J. Organomet. Chem.* **182**, 477 (1979).
3. J. J. Eisch, *Organomet. Synth.* **2**, 120 (1981); *cf.* A. N. Nesmeyanov, A. E. Borisov and N. A. Vol'kanau, *Izv. Akad. Nauk SSSR Otdel. Khim. Nauk* 992 (1954).
4. C. Tessier-Youngs and O. T. Beachley, *Inorg. Synth.* **24**, 92 (1986).
5. H. H. Karsch, A. Appelt, J. Riede and G. Müller, *Organometallics* **6**, 316 (1987).
6. R. Uson and A. Laguna, *Inorg. Synth.* **21**, 72 (1982).
7. J. J. Eisch, *Organomet. Synth.* **2**, 154 (1981); *cf.* H. Gilman and R. G. Jones, *J. Am. Chem. Soc.* **61**, 1513 (1939) and **62**, 2537 (1940).
8. J. J. Eisch, *Organomet. Synth.* **2**, 167 (1981); *cf.* O. H. Johnson and W. H. Nebergall, *J. Am. Chem. Soc.* **71**, 1720 (1949).
9. S. C. Cohen and A. G. Massey, *Organomet. Synth.* **3**, 563 (1986).
10. J. J. Eisch, *Organomet. Synth.* **2**, 177 (1981); *cf.* J. J. Eisch, N. K. Hota and S. Kozima, *J. Am. Chem. Soc.* **91**, 4535 (1969).

REFERENCES

11. D. Hellwinkel, *Organomet. Synth.* **3**, 615 (1986); see also *Ibid.* p. 628 for a related antimony compound.
12. J. A. J. Jarvis, R. Pearce and M. F. Lappert, *J. Chem. Soc. Dalton Trans.* 999 (1977).
13. D. E. Bergbreiter, S. S. Shimazu and T. J. Lynch, *Organomet. Synth.* **3**, 346 (1986).
14. D. E. Fenton, A. G. Massey, A. J. Park and V. B. Smith, *Organomet. Synth.* **3**, 343 (1986).
15. R. Uson and A. Laguna, *Organomet. Synth.* **3**, 322 (1986).
16. G. van Koten, C. A. Schaap, J. T. B. H. Jastrzebski and J. G. Noltes, *J. Organomet. Chem.* **186**, 427 (1980).
17. K. Clauss, *Justus Liebigs Ann. Chem.* **711**, 19 (1968).
18. M. D. Rausch and H. B. Gordon, *J. Organomet. Chem.* **74**, 85 (1974).
19. W. Mowat, A. Shortland, G. Yagupsky, N. J. Hill, M. Yagupsky and G. Wilkinson, *J. Chem. Soc. Dalton Trans.* 533 (1972).
20. K. L. Brandenburg and H. B. Abrahamson, *J. Organomet. Chem.* **293**, 371 (1986).
21. J. Müller and H. Menig, *J. Organomet. Chem.* **191**, 303 (1980).
22. H. J. Jakobsen, E. H. Larsen and S.-O. Lawesson, *Tetrahedron* **19**, 1867 (1963).
23. J. J. Eisch, *Organomet. Synth.* **2**, 110 (1981).
24. L. F. Fieser, *Reagents for Organic Synthesis*, Vol. 1, p. 629, Wiley, New York.
25. A. D. Baxter, F. Binns, T. Javed, S. M. Roberts, P. Sadler, F. Scheinmann, B. J. Wakefield, M. Lynch and R. F. Newton, *J. Chem. Soc. Perkin Trans. 1* 889 (1986).
26. See Section 3.2, Ref. (*25*).
27. See Section 3.2, Refs. (*26*), (*27*) and (*39*).
28. C. Petrier, J. L. Luche and C. Dupuy, *Tetrahedron Lett.* **25**, 3463 (1984).
29. C. Petrier, J. C. de S. Barbosa, C. Dupuy and J. L. Luche, *J. Org. Chem.* **50**, 5761 (1985).
30. F. Huet, J. Michael, C. Bernardon and E. Henry-Basch, *C. R. Acad. Sci. Paris* **262C**, 1328 (1966).
31. See Section 6.1.2, Refs (*3*) and (*4*), and Section 8.1, Ref. (*3*); see also S. H. Bertz, C. P. Gibson and G. B. Dabbagh, *Tetrahedron Lett.* **28**, 4251 (1987).
32. See Section 6.1.1, Refs (*9*) and (*10*).
33. T. Kauffmann, K. Abel, W. Bonrath, M. Kolb, T. Möller, C. Pahde, S. Raedecker, M. Robert, M. Wensing and B. Wichmann, *Tetrahedron Lett.* **27**, 5351 (1986).
34. T. Kauffman, T. Abel, C. Beirich, G. Kieper, C. Pahde, M. Schreer, E. Toliopoulos and R. Wieschollek, *Tetrahedron Lett.* **27**, 5355 (1986).
35. See Section 6.1.1, Ref. (*11*).
36. J. R. Long, *Aldrichimica Acta* **18**, 87 (1985).
37. T. Inamoto and Y. Sugiura, *J. Organomet. Chem.* **285**, C21 (1985).
38. L. A. Paquette and K. S. Learn, *J. Am. Chem. Soc.* **108**, 7873 (1986).
39. S. E. Denmark, T. Weber and D. W. Piotrowski, *J. Am. Chem. Soc.* **109**, 2224 (1987).
40. C. R. Johnson and B. D. Tait, *J. Org. Chem.* **52**, 281 (1987).
41. See Section 8.1, Ref. (*4*).
42. T. Kauffman and F. Steinseifer, *Chem. Ber.* **118**, 1031 (1985).
43. H. A. Brune, R. Hess and G. Schmidtberg, *J. Organomet. Chem.* **303**, 429 (1986).

—16—
Applications of Elimination Reactions of Organolithium Compounds; Arynes, Carbenes, Ylides, Ring Opening of Heterocycles

In this chapter are collected examples of reactions that can be represented on paper as involving elimination of a lithium cation and an anion such as a halide. The emphasis is on the outcome of the reactions, and such topics as the dividing lines between bimolecular base-induced eliminations (E2) and eliminations involving a discrete organolithium intermediate (related to E1$_{cb}$), or whether free arynes or carbenes are intermediates, will not be discussed in any detail. Nevertheless, it must be emphasized that it is often an oversimplification to regard the organolithium compound merely as a strong base, and to ignore the possible role of the lithium salt eliminated.

16.1. ORGANOLITHIUM COMPOUNDS AS PRECURSORS OF ARYNES

The elimination-addition (benzyne) mechanism of aromatic substitution was established by classical studies of reactions of organolithium compounds with aryl halides (*1*). This earlier work involved deprotonation, and any intermediate *o*-halogenoaryllithium compound was rarely observed. However, the speed of the lithium–halogen exchange reaction even at low temperatures enabled solutions of *o*-halogenoaryllithium compounds to be prepared, and reactions carried out with them (see Section 3.1.3). The solutions could then be warmed in the presence of a co-reagent to trap the aryne. The reaction of an aryl halide with a strong non-nucleophilic base remains a useful alternative method for generating arynes and hetarynes, which in the case of aryl bromides avoids complications caused by metal–halogen exchange. LiTMP (*2*) and LDA (*3*) are particularly effective, and with these reagents it is often possible to intercept the intermediate *o*-halogenoaryllithium compounds.

TABLE 16.1
Cycloadducts of Arynes Formed from Organolithium Compounds

Organolithium compound	Trapping reagent	Product (yield %)	Ref.
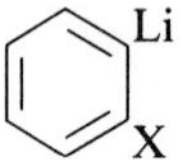 o-C$_6$H$_4$(Li)(X), X = F, Cl, Br	 furan	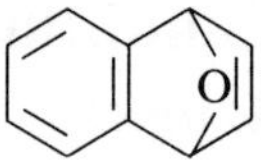 1,4-epoxy-1,4-dihydronaphthalene (up to 84)	(4)[a]
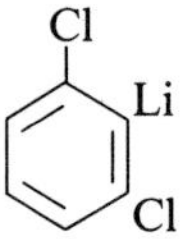 2,3-dichlorophenyllithium (Cl, Li, Cl)	 N-methylpyrrole	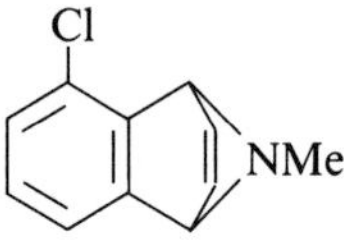 5-chloro-1,4-dihydro-1,4-iminonaphthalene NMe (38)[b]	(5)
C$_6$Cl$_5$Li	C$_6$H$_6$	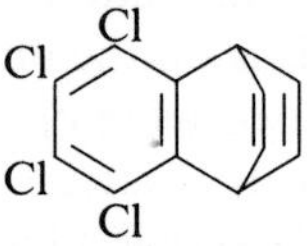 tetrachlorobenzobicyclo adduct (58–60)	(6)
 2-chloro-3-lithiopyridine	 furan	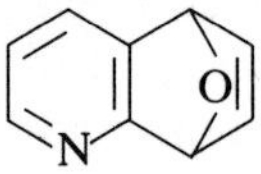 pyridine-fused epoxy adduct (31)[c]	(7)

[a] This paper contains a thorough discussion of the factors required to obtain good yields. [b] A small proportion of products derived from 4-chlorobenzyne via 2,4-dichlorophenyllithium may also be formed (8). [c] Yield is probably higher, but product is unstable.

Some o-halogenoaryllithium compounds have been tabulated, as have their aryne-forming reactions (see General Ref. A). Some well-described examples of cycloaddition reactions of arynes derived from organolithium intermediates are listed in Table 16.1, and full details are given for one, involving the relatively stable pentachlorophenyllithium.

Generation and trapping of tetrachlorobenzyne;
5,8-epoxy-1,2,3,4-tetrachloro-5,8-dihydronaphthalene (9)

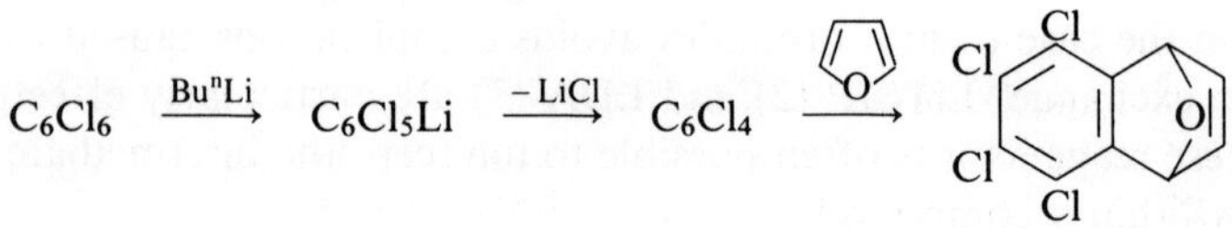

$$C_6Cl_6 \xrightarrow{Bu^nLi} C_6Cl_5Li \xrightarrow{-LiCl} C_6Cl_4 \xrightarrow{\text{furan}} \text{5,8-epoxy-1,2,3,4-tetrachloro-5,8-dihydronaphthalene}$$

16.2. CARBENOID REACTIONS 169

A solution of pentachlorophenyllithium is prepared as described on p. 28 from hexachlorobenzene (6.0 g, 21.0 mmol) in ether (150 ml). The solution is cooled to $-78°$ and stirred as freshly distilled furan (22 ml) is added dropwise during 1 h. The mixture is allowed to warm slowly to room temperature and stirred at room temperature for *ca.* 40 h. The solvents are evaporated under reduced pressure. Chromatography of the dark-tan solid residue (alumina, 6% water added, 1:1 hexane–benzene[a]) gives a little hexachlorobenzene followed by a pale-yellow solid. Sublimation of the latter (90° *in vacuo*) gives 5,8-epoxy-1,2,3,4-tetrachloro-5,8-dihydronaphthalene (82%), m.p. 119.5–120.5°.

[a] Hazard: an alternative solvent would be preferable.

References

1. R. W. Hoffmann, *Dehydrobenzene and Cycloalkynes*, Academic Press, New York, 1967.
2. D. J. Pollart and B. Rickborn, *J. Org. Chem.* **52**, 792 (1987) and references therein.
3. M. E. Jung and G. T. Lowen, *Tetrahedron Lett.* **27**, 5319 (1986).
4. H. Gilman and R. D. Gorsich, *J. Am. Chem. Soc.* **79**, 2625 (1957).
5. L. A. Motyka, *Tetrahedron Lett.* **26**, 2827 (1985).
6. See Section 3.1.3, Ref. (*18*).
7. J. D. Cook and B. J. Wakefield, *J. Chem. Soc. (C)* 1973 (1969).
8. S. Seki, T. Morinaga, H. Kikuchi, T. Mitsuhashi, G. Yamamoto and M. Oki, *Bull. Chem. Soc. Jpn* **54**, 1465 (1981).
9. G. A. Moser, F. E. Tibbetts and M. D. Rausch, *Organomet. Chem. Synth.* **1**, 99 (1970/1).

16.2. ORGANOLITHIUM CARBENOID REACTIONS

Aspects of the carbenoid chemistry of α-halogenoorganolithium compounds (or the corresponding carbanions) are discussed in General Refs A and D(ii), as well as in general works on the chemistry of carbenes (*1*).

16.2.1. Synthesis of Cyclopropanes and Oxiranes

In the examples reviewed here the products are formally adducts of carbenes to carbon–carbon or carbon–oxygen multiple bonds. In these overall "cycloadditions" the reaction could follow several pathways:

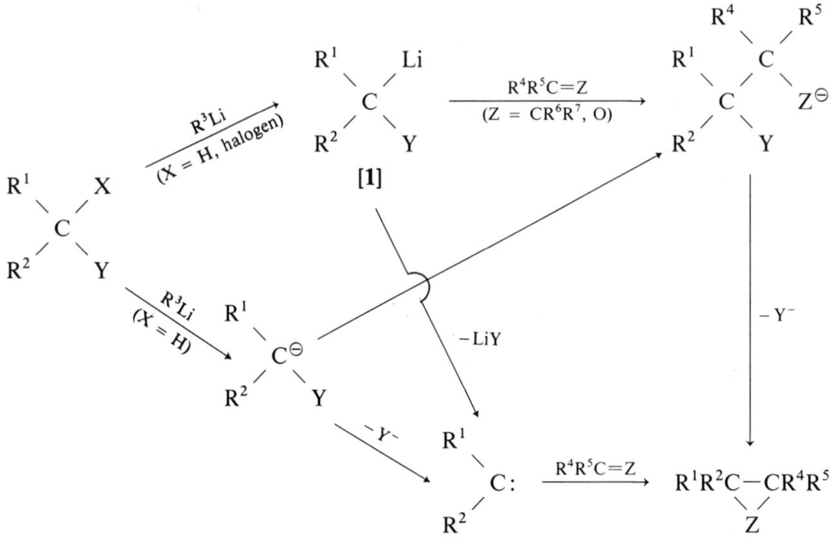

16. APPLICATIONS OF ELIMINATION REACTIONS

Even this scheme is an oversimplification, and mechanistic details will not be discussed, though the broad generalization can be made that the additions to alkenes tend to be closer to true carbene reactions, whereas the reactions with carbonyl compounds usually follow addition–elimination sequences. Some examples where organolithium intermediates ·1· have been generated and intercepted at low temperatures have been described earlier (see e.g. pp. 80, 81, 92).

Some carbenoid cycloadditions to alkenes and alkynes are listed in Table 16.2, and others are tabulated in General Refs A and D(ii). Syntheses of oxiranes by reactions of lithium carbenoids with carbonyl compounds are listed in Table 16.3.

Carbenoid cycloaddition via dibromofluoromethyllithium;
1-bromo-1-fluoro-2,2,3,3-tetramethylcyclopropane (10)

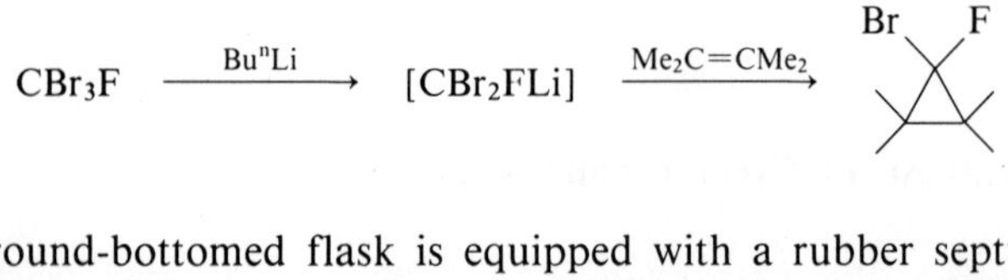

$$CBr_3F \xrightarrow{Bu^nLi} [CBr_2FLi] \xrightarrow{Me_2C=CMe_2}$$

A 50 ml round-bottomed flask is equipped with a rubber septum and a mechanical stirrer,^{a} and furnished with an atmosphere of argon. The flask is cooled in an ethanol slush bath ($-116°$) and charged with THF (15 ml), tribromofluoromethane (2.707 g, 10 mmol) and 2,3-dimethyl-2-butene

16.2. CARBENOID REACTIONS

TABLE 16.2

Synthesis of Cyclopropanes by Organolithium Carbenoid Cycloadditions

Carbenoid	Substrate	Product (yield %)		Ref.
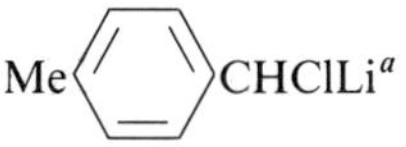 Me-C₆H₄-CHClLi[a]	$CH_2=CHOEt$	Me-C₆H₄-cyclopropyl-OEt	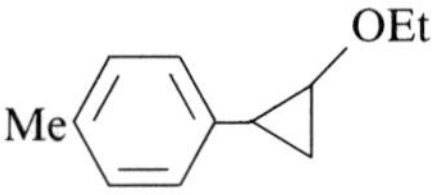 (75–80)	(2)
$CH_2=CHCCl_2Li$[a]	$Me_2C=CH_2$	Cl-cyclopropyl-vinyl	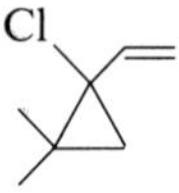 (42)	(3)
PhCHClLi	$MeC\equiv CMe$	(43)[b] cyclopropene isomers with Ph		(4)
$ClCH_2CH_2OCHClLi$	 cyclohexene	bicyclic-OCH₂CH₂Cl	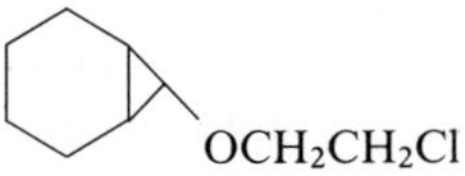 (40–56)	(5)

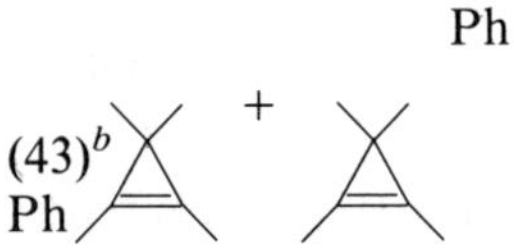

[a] Intermediate generated by deprotonation by LiTMP; may not be true organolithium compound. [b] Mixture; by reaction of initial adduct [1] with MeLi: Ph-cyclopropene [1]

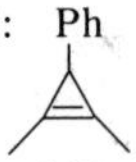

TABLE 16.3

Synthesis of Oxiranes by Reactions of Organolithium Carbenoids with Carbonyl Compounds

Organolithium compound	Carbonyl compound	Product (yield %)		Ref.[a]
CH_2BrLi	PhCOMe	Ph,Me-oxirane	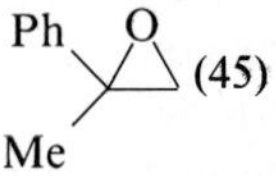 (45)	(6)
MeCHBrLi	n-C_7H_{15}CHO	C_7H_{15},Me-oxirane	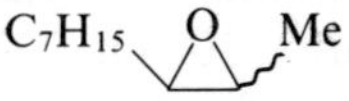 (80)	(7)
$CH_2=CCClLi$ / Me	MeCOCHMe₂	Me,Me,Me₂CH,CH=CH₂-oxirane	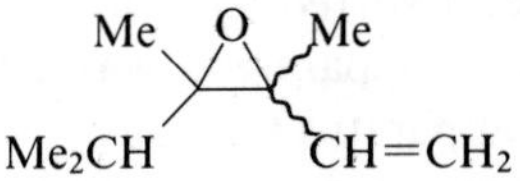 (74)	(8)

[a] Each of the references cited describes several examples. See also Ref. (9).

(0.842 g, 10 mmol). The mixture is stirred vigorously as freshly prepared n-butyllithium[b] (in hexane, 10 mmol) is added slowly by syringe during ca. 20 min. Stirring and cooling is continued for 15 min. Water (1 ml) is added and the mixture is allowed to warm to room temperature. The organic phase contains 1-bromo-1-fluoro-2,2,3,3-tetramethylcyclopropane (73%).[c]

[a] The reaction mixture becomes viscous at the low temperatures used, and efficient stirring is important. With magnetic stirring the yield was reduced to 53%.
[b] When commercial butyllithium was used the yield was reduced to 61%.
[c] Analysis by ^{19}F NMR (benzotrifluoride as internal standard). The product is unstable.

16.2.2. Organolithium Carbenoid Insertion Reactions

The following example (11) is representative of a variety of such reactions, both inter- and intramolecular. References to other examples are given in General Refs A and D(ii). In this example the organolithium carbenoid is prepared from a *gem*-dihalide by lithium–halogen exchange, as is usually necessary. Recently, a similar reaction has been carried out satisfactorily by reaction of the dihalide with lithium metal under ultrasonic irradiation (12).

1,6-Dimethyltricyclo[4.1.0.0^{2,7}]hept-3-ene

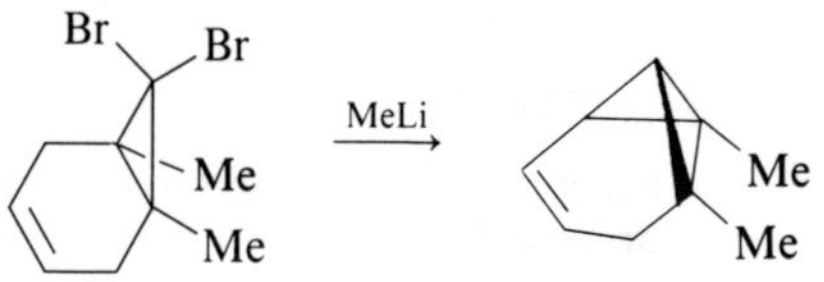

A solution of 7,7-dibromo-1,6-dimethylbicyclo[4.1.0]hept-3-ene[a] (20.95 g, 75 mmol) in ether (500 ml) is placed in a 1 l three-necked flask equipped with a magnetic stirrer, a reflux condenser and a dropping funnel, and furnished with an atmosphere of nitrogen. The solution is cooled (ice bath) and stirred as a mixture of methyllithium (ca. 1.6 M in ether, 80 mmol) and ether (70 ml) is added dropwise during 30 min. The mixture is allowed to warm to room temperature and stirred for 1 h. Water (100 ml) is added cautiously. The organic layer is separated, washed with water (3 × 100 ml) and dried (Na$_2$SO$_4$). In apparatus previously washed in base and dried, the solution is carefully concentrated by slow distillation through a 40 cm Vigreux column (bath temperature not above 60°). Also in base-washed apparatus, the

16.2. CARBENOID REACTIONS

residue is distilled through a short unpacked column, the fraction b.p. 48–49°/23 mmHg being collected as 1,6-dimethyltricyclo[4.1.0.0^{2,7}]hept-3-ene (4.2–4.4 g, 46–49%).

^{a} The preparation of this compound is also described (*11*).

16.2.3. Organolithium Carbenoid Rearrangement Reactions

As in the case of insertion reactions, a variety of rearrangements of organolithium carbenoids is possible. Many such reactions are of academic interest rather than practical value. However, rearrangements of 1-halogeno-1-lithioalkenes are a useful route to unsymmetrical diarylacetylenes (see General Ref. A):*

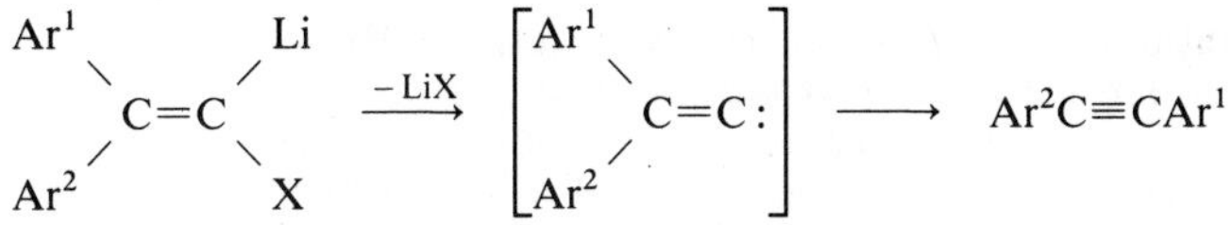

$$\underset{Ar^2}{\overset{Ar^1}{\diagdown}}C=C\underset{X}{\overset{Li}{\diagup}} \quad \xrightarrow{-LiX} \quad \left[\underset{Ar^2}{\overset{Ar^1}{\diagdown}}C=C:\right] \quad \longrightarrow \quad Ar^2C\equiv CAr^1$$

Furthermore, rearrangements of β-oxidocarbenoids are the key step in a general method for ring expansion of cyclic ketones:

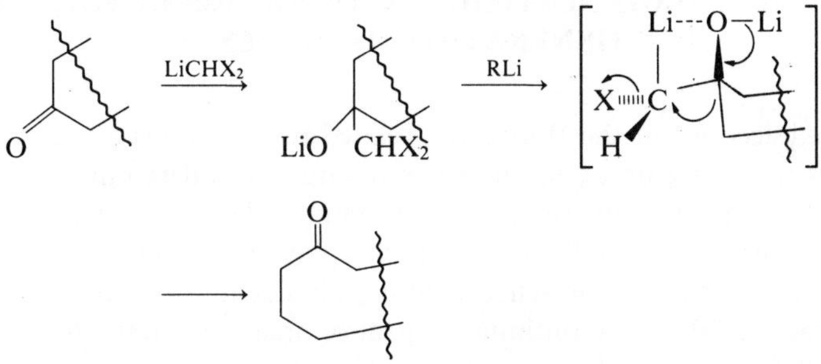

Reactions of this latter type, whose generality was demonstrated notably by Nozaki *et al.* (*13*), can show good regioselectivity and have proved useful in the synthesis of natural products. The following is a recent well-described example (*14*):

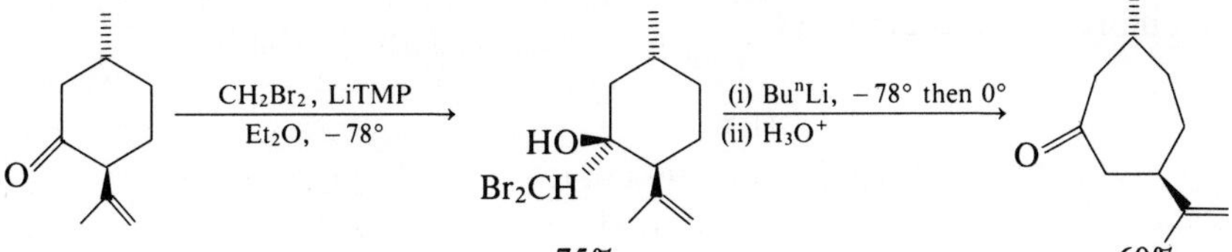

* A carbene intermediate is shown for clarity, but elimination and rearrangement may well be concerted.

References

1. W. Kirmse, *Carbene Chemistry*, 2nd edn, Academic Press, New York, 1971.
2. C. M. Dougherty and R. A. Olofson, *Org. Synth.* **58**, 37 (1978).
3. R. A. Moss and R. C. Munjal, *Synthesis* 425 (1979).
4. L. E. Friedric and R. A. Fiato, *Synthesis* 611 (1973).
5. U. Schöllkopf, J. Paust and M. R. Patsch, *Org. Synth.* **49**, 86 (1969); see also P. v. R. Schleyer, W. F. Sliwinski, G. W. Van Dine, U. Schöllkopf, J. Paust and K. Fellenberger, *J. Am. Chem. Soc.* **94**, 125 (1972).
6. G. Cainelli, A. Umani-Ronchi, F. Bertini, P. Grasselli and G. Zubiani, *Tetrahedron* **27**, 6109 (1971).
7. G. Cainelli, N. Tangari and A. Umani-Ronchi, *Tetrahedron* **28**, 3009 (1972).
8. B. Mauzé, *J. Organomet. Chem.* **170**, 265 (1979).
9. R. Tarhouni, B. Kirschleger and J. Villieras, *J. Organomet. Chem.* **272**, C1 (1984).
10. D. J. Burton and J. L. Hahnfeld, *J. Org. Chem.* **42**, 830 (1977).
11. R. T. Taylor and L. A. Paquette, *Org. Synth.* **61**, 39 (1981).
12. L. Xu, F. Tao and T. Yu, *Tetrahedron Lett.* **26**, 4231 (1985).
13. H. Taguchi, H. Yamamoto and H. Nozaki, *Bull. Chem. Soc. Jpn* **50**, 1588 and 1592 (1977).
14. R. Bandouy, J. Gore and L. Ruest, *Tetrahedron* **43**, 1009 (1987).

16.3. ORGANOLITHIUM COMPOUNDS IN THE GENERATION OF YLIDES

A detailed account of the Wittig and related reactions, even if confined to examples involving organolithium compounds, would require a complete volume. In some cases an organolithium compound is used merely as a strong base for generating an ylide, but in others it plays a more intimate role in the reaction. Moreover, the presence of lithium halide in the solution (either as a constituent of the organolithium reagent or arising from the formation of the ylide) can affect the reaction, and in particular its stereochemical outcome (see below).

A full account of the "classical" Wittig reaction, with examples of experimental procedures, is given by Maercker (*1*), and the preparation of methylenecyclohexane is described in *Organic Syntheses (2)*. The following is a more recent example (*3*):

16.3. GENERATION OF YLIDES

Some alkene-forming reactions related to the Wittig reaction, but not involving true ylides, are noted in Section 6.1.3.

The control of the stereochemistry of reactions of ylides with aldehydes is of great importance. In general, unstabilized ylides, in the absence of soluble metal salts (particularly lithium salts), tend to give Z-alkenes. However, in the presence of lithium salts the proportion of E-alkene in the product is increased, particularly at higher concentrations. The influence of such factors on the stereochemistry of the Wittig reaction, and the explanation for these influences, have been thoroughly investigated by Maryanoff et al. (4, 5).

If an E-alkene is desired then the procedure of Schlosser et al. is adopted, as in the following example (6, 7). The essential feature of this procedure is the use of an excess of organolithium compound, which metallates the Wittig intermediate (betaine/oxaphosphetane) to give a lithio-β-oxidoylide, which undergoes equilibration. Protonation then gives the more thermodynamically stable betaine/oxaphosphetane, and subsequent elimination leads to the E-alkene. In the Schlosser procedure the last two stages are accomplished by means of a potassium t-butoxide–t-butanol complex.

Selective formation of an E-*alkene by a modified Wittig reaction;*
E-*oct-2-ene*

$$Ph_3\overset{\oplus}{P}CH_2CH_3 \ Br^{\ominus} \quad \begin{array}{l}\text{(i) PhLi}\\ \text{(ii) n-}C_5H_{11}CHO\\ \text{(iii) PhLi}\\ \text{(iv) Bu}^tOK\end{array} \quad \underset{C_5H_{11}}{\diagup\!=\!\diagdown}^{CH_3}$$

A mixture of ethyltriphenylphosphonium bromide (11.14 g, 30 mmol), THF (60 ml) and ether (30 ml) is stirred and cooled to $-78°$ under an atmosphere of nitrogen. Phenyllithium (ca. 0.9 M in ether,[a] 30 mmol) is added dropwise to the stirred solution. After a further 30 min stirring a solution of hexanal (3.0 g, 30 mmol) in ether is added, followed, after the colour of the solution fades (ca. 10 min), by more phenyllithium solution (30 mmol). Stirring is continued as the temperature of the reaction mixture is allowed to rise to $-30°$. Potassium t-butoxide–t-butanol complex (45 mmol) is added and stirring is continued for 1 h. The reaction mixture is centrifuged at ca. 5000 rpm. The clear solution is decanted, washed with water until neutral, and dried. The solvents are distilled through a packed column and a fraction b.p. 80–120° (4 g) is collected. This distillate contains (as shown by gas chromatography) oct-2-ene ($E:Z$ ratio 99:1; 69% yield based on the phosphonium salt) together with ether, THF, benzene and t-butanol.

[a] The solution was prepared from bromobenzene and lithium in ether and was 1.2 M in lithium bromide (6).

References

1. A. Maercker, *Org. React.* **14**, 270 (1965).
2. G. Wittig and U. Schöllkopf, *Org. Synth. Coll. Vol.* **5**, 752 (1973).
3. E. J. Leopold, *Org. Synth.* **64**, 164 (1986).
4. A. B. Reitz, S. O. Nortey, A. D. Jordan, M. S. Mutter and B. E. Maryanoff, *J. Org. Chem.* **51**, 3302 (1986).
5. B. E. Maryanoff, A. B. Reitz, M. S. Mutter, R. R. Inners, H. R. Almond, R. R. Whittle and R. A. Olofson, *J. Am. Chem. Soc.* **108**, 7664 (1986).
6. M. Schlosser and K. F. Christmann, *Liebigs Ann. Chem.* **708**, 1 (1967).
7. M. Schlosser, K. F. Christmann and A. Piskala, *Chem. Ber.* **103**, 2814 (1970); see also B. E. Maryanoff, A. B. Reitz and B. A. Duhl-Emswiler, *J. Am. Chem. Soc.* **107**, 217 (1985).

16.4. SYNTHESES VIA RING OPENING OF LITHIATED HETEROCYCLES

Several types of reaction are known where lithiated heterocycles undergo elimination leading to ring opening, and in a number of cases the products of such reactions are useful in themselves or as intermediates for further reactions. The most widely explored type of ring opening is via β-elimination:

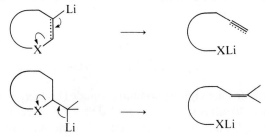

However, some α-eliminations are also known:

Some examples of the various types of ring opening are listed in Table 16.4, and a reaction involving β-elimination is fully described below. Other examples are reviewed in general Refs A, D(ii) and (*10a*).

The reaction under the heading "cycloreversion" in Table 16.4 is the most important of a small group of reactions characterized as 1,3-anionic cycloreversions (*11*). This cleavage of THF by organolithium compounds via

TABLE 16.4

Syntheses via Ring Opening of Lithiated Heterocycles

Lithiated heterocyclic compound	Ring-opened intermediate	Product isolated (yield %) (further reagent added)	Ref.
β-elimination (ring lithium)			
Ph-furan-Ph (2-Li)	Ph-C(OLi)=C≡-Ph	PhCOCH$_2$C≡CPh (75) (H$_3$O$^+$)	(1)
benzofuran (2-Li)	C$_6$H$_4$(OLi)-C≡CH	C$_6$H$_4$(OH)-C≡CH (67) (H$_3$O$^+$)	(2)
Me-thiophene-Me (3-Li)	Me-C(SLi)=C≡-Me	Me-C(SMe)=C≡-Me (63) (Me$_2$SO$_4$)	(3)
benzothiophene (2-Li)	C$_6$H$_4$(SLi)-C≡CH	C$_6$H$_4$(SMe)-C≡CMe (42) (Me$_2$SO$_4$)	(4)
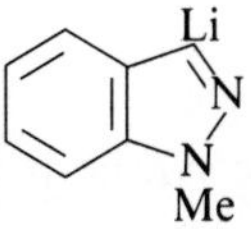 N-methylindazole (3-Li)	 C$_6$H$_4$(NLiMe)-CN	 C$_6$H$_4$(NHMe)-COPh (50) ((i) PhLi, (ii) H$_3$O$^+$)	(5)
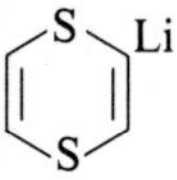 1,3-dithiine (2-Li)	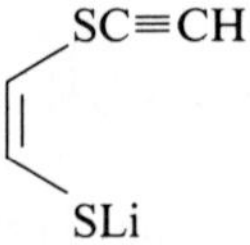 SC≡CH / CH$_2$=CH-SLi	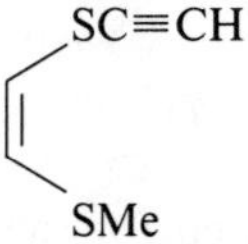 SC≡CH / CH$_2$=CH-SMe (>90) (MeI)	(6)

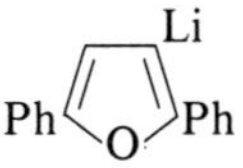

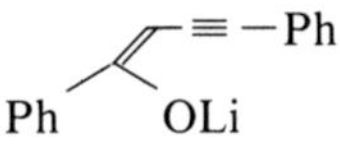

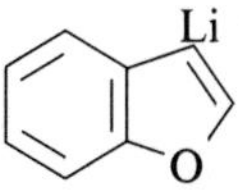

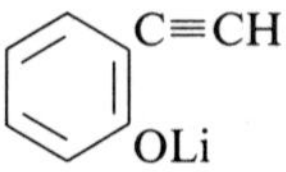

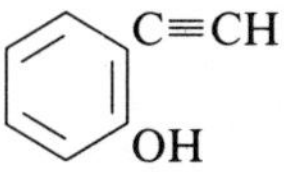

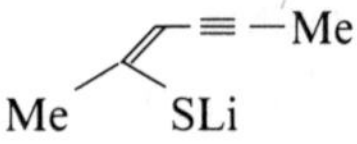

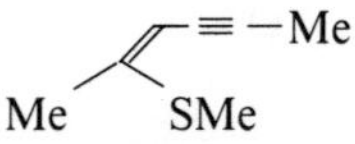

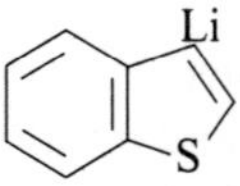

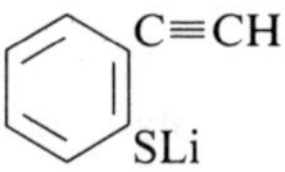

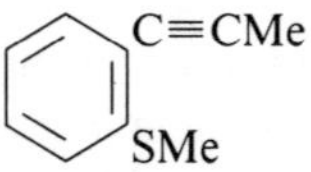

16. APPLICATIONS OF ELIMINATION REACTIONS

TABLE 16.4—continued

β-elimination (exocyclic lithium)

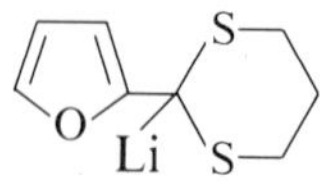

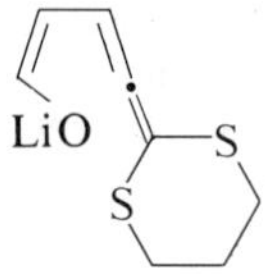

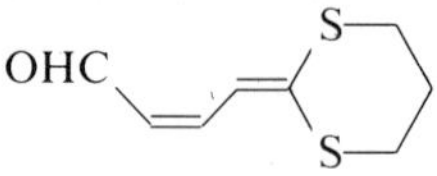

$$\text{OHC}\diagup\!\!\diagdown\text{(quant.)} \quad (7)$$

((i) Me$_3$SiCl, (ii) H$_2$O)

α-elimination

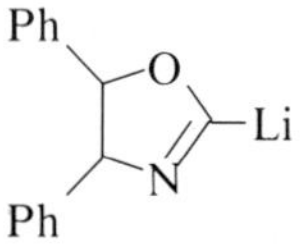

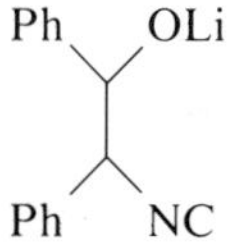

(a)

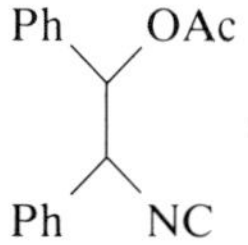

(81) (Ac$_2$O)   (8)

Cycloreversion

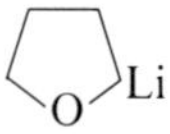

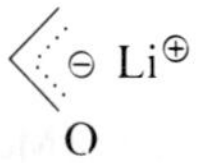

(83)   (9)

OSiMe$_3$
(Me$_3$SiCl)

---

a The ring-opened product and the lithiated heterocycle are in tautomeric equilibrium. With other reagents some 2-substituted oxazoline is obtained. Oxazoles behave similarly (*10*).

α-metallation is a nuisance when THF is merely required as a solvent, but it is a useful means for cleanly generating the enolate of ethanal, and it has also been used to liberate ethene in the presence of a co-reactant (*9, 12, 13*):

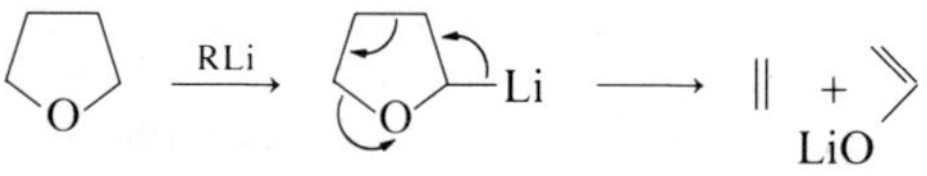

Analogous reactions of ketals have been used for a stereospecific synthesis of alkenes (*14*):

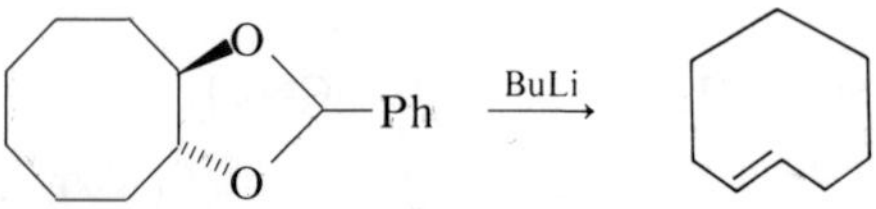

They have also been used for the generation of the enolate of crotonaldehyde (*15*) and of lithium ethynolate (*16*). The cleavage of ethers by organolithium compounds has recently been reviewed (*17*).

Synthesis of a selenoenyne by ring opening of lithioselenophene (18)

$$\underset{Se}{\overset{Br}{\underset{Me}{\bigvee}}}Et \xrightarrow{EtLi^a} \underset{Se}{\overset{Li}{\underset{Me}{\bigvee}}}Et \longrightarrow \underset{Me}{\overset{\equiv -Et}{\underset{SeLi}{\bigwedge}}} \xrightarrow{EtBr} \underset{Me}{\overset{\equiv -Et}{\underset{SeEt}{\bigwedge}}}$$

3-Bromo-2-ethyl-5-methylselenophene (2.52 g, 10 mmol) is dissolved in ether (50 ml). The solution is stirred as ethyllithium [a] (ca. 0.6 M in ether, 10 mmol) is added, followed by bromoethane (5.45 g, 50 mmol). The solution is washed with water and dried, and the solvent is evaporated. Paraffin oil is added to the residue and the product is distilled, a fraction b.p. 108–109°/ 12 mmHg being (Z)-2-(ethylseleno)hept-2-en-4-yne (1.1 g, 54%).

[a] The alkyllithium compound has to correspond to the final alkylating reagent, since alkyl halide is produced by the lithium–halogen exchange reaction.

References

1. T. L. Gilchrist and D. J. Pearson, *J. Chem. Soc. Perkins Trans. 1* 989 (1976).
2. H. Gilman and D. S. Melstrom, *J. Am. Chem. Soc.* **70**, 1655 (1948).
3. S. Gronowitz and T. Frejd, *Acta Chem. Scand.* **B30**, 287 (1976); see also T. Frejd, *J. Heterocycl. Chem.* **14**, 1085 (1977).
4. R. P. Dickinson and B. Iddon, *J. Chem. Soc. (C)* 3447 (1971).
5. B. A. Tertov, V. V. Bessonov and P. P. Onishchenko, *J. Org. Chem. USSR (Engl. Transl.)* **10**, 2634 (1974).
6. M. Schoufs, J. Meyer, P. Vermeer and L. Brandsma, *Recl. Trav. Chim. Pays-Bas* **96**, 259 (1977).
7. M. J. Taschner and G. A. Kraus, *J. Org. Chem.* **43**, 4235 (1978).
8. U. Schöllkopf, F. Gerhart, I. Hoppe, R. Harms, K. Hantke, K.-H. Scheunemann, E. Eilers and E. Blume, *Justus Liebigs Ann. Chem.* 183 (1976).
9. M. E. Jung and R. B. Blum, *Tetrahedron Lett.* 3791 (1977).
10. L. N. Pridgen and S. C. Shilcrat, *Synthesis* 1048 (1984) and Refs. therein.
10a. T. Gilchrist, *Adv. Heterocycl. Chem.* **41**, 41 (1987).
11. G. Bianchi, C. De Micheli and R. Gandolfi, *Angew. Chem. Int. Ed. Engl.* **18**, 721 (1979).
12. R. B. Bates, L. M. Kroposki and D. E. Potter, *J. Org. Chem.* **37**, 560 (1972); see also A. Maercker and W. Theysohn, *Justus Liebigs Ann. Chem.* **747**, 70 (1971).
13. K. Kamata and M. Terashima, *Heterocycles* **14**, 205 (1980).
14. J. N. Hines, M. J. Peagram, E. J. Thomas, and G. H. Whitham, *J. Chem. Soc. Perkin Trans. 1* 2332 (1973).
15. G. Demailly, J. B. Ousset and C. Mioskowski, *Tetrahedron Lett.* **25**, 4647 (1984).
16. B. L. Groh, G. R. Magrum and T. J. Barton, *J. Am. Chem. Soc.* **109**, 7568 (1987).
17. A. Meaercker, *Angew. Chem. Int. Ed. Engl.* **26**, 972 (1987).
18. S. Gronowitz and T. Frejd, *Acta Chem. Scand.* **B30**, 313 (1976).

Index of Compounds and Methods

All organolithium compounds mentioned in the text and tables are indexed, but for methyllithium, n-butyllithium, s-butyllithium, t-butyllithium and phenyllithium (and also LDA) only selected entries are included. Where applicable, the latest entry numbers in the *Dictionary of Organometallic Compounds* (General Ref. E and supplements) are given in parentheses.

In general, substrates for reactions of organolithium compounds are indexed under the names of the class of compound, rather than as individual compounds, and references to routine uses of common substrate classes such as carbonyl compounds, alkyl halides and trimethylsilyl chloride are excluded.

ff denotes following folios.

A

Acetals, 107
Acetylcyclohexane, 93
Acyl halides, 76ff
Acyllithium compounds, 95–97
1-Adamantyllithium (Li-10033), 25
Aldehydes
   reactions with organolithium compounds, 67ff, 96
   synthesis of, 76, 82ff, 93
Aldol reaction, 37
Alkenes
   addition of organolithium compounds, 53ff
   metallation, 36, 39
Alkyl halides, 21ff, 107ff, 143ff
Alkynes
   addition of organolithium compounds, 53
   metallation, 33, 40
Allyl alcohol, 54

Allyl ethers
   cleavage of, 49
Allyllithium (Li-00033), 22, 39, 46, 47, 70
Allylpotassium (K-20001), 39
Allyl 2,4,6-trimethylbenzoates, 49
Amides (organic; *see also* Lithium amides)
   metal–halogen exchange of halogenated, 28
   metallation of, 33–35, 39
   synthesis of, 83
   synthesis of aldehydes and lactones from, 82ff
   'vinylogous', 86
Amines
   N-deprotonation of, 119–120
   synthesis of, 125ff
2-Amino-3,3'-bithienyl, 127
Anhydrides, 76ff
Arenes
   addition of organolithium compounds, 53
   tricarbonylchromium complexes, 53, 55
Argon, 9, 23, 26

181

Arynes, 167–169
Association of organolithium compounds, 3–4
5-Aza-6-lithio-2,2,4,4,9-pentamethyl non-5-ene (Li-30060), 66, 121
5-Aza-6-lithio-2,2,4,4-tetramethyldec-5-ene, 114
Azides, 127
2-Azido-3,3'-bithienyl, 127

B

Barbier-type syntheses, 69, 70, 80, 83, 89
Benzene, 5
Benzo[b]thiophene-2(3H)-one, 151
Benzylalcoholtricarbonyl chromium, 53
Benzyl ethers
  cleavage of, 47
N-Benzylidenebenzylamine as indicator, 18
1-Benzyl-5-lithio-3-phenyl-1,2,4-triazole, 156
1-Benzyl-5-lithiothieno[3,2-e]indole, 77
Benzyllithium (Li-30024), 38
Biphenyl-4-ylmethanol as indicator, 18
2,2'-Biquinoline as indicator, 18
Bis(dimethylphosphino)methyllithium, 160
Bis(phenylthio)methyllithium (Li-10042), 40, 72
Bis(trimethylsilyl)peroxide, 132
cis-Bis(4-trimethylsilyl)phenyl)bis(triphenylphosphine)platinum, 164
Borate esters, 150–151
2-Bornyllithium, 25
Boron halides, 149–151
Boron trifluoride, 113, 116, 150
Bromine, 143, 144
Bromobenzene
  reaction with lithium, 24
1-Bromoethyllithium, 171
1-Bromo-1-fluoro-2,2,3,3-tetramethylcyclopropane, 170
7-Bromo-7-lithio-1,6-dimethylbicyclo[4.1.0]hept-3-ene, 172
7-exo-Bromo-7-endo-lithio-2-oxabicyclo[4.1.0]heptane, 124
Bromomethyllithium (Li-10001), 171, 173
1-Bromopentyllithium, 30
2-Bromophenyllithium (Li-10015), 168
2-Bromotetrafluorophenyllithium, 160
2-Buten-2-yllithium, 25

2-t-Butoxy-5-lithiothiophene, 111
t-Butyl chloride
  reaction with lithium, 23
t-Butylcyanide, 64
trans-2-Butylcyclohexanol, 116
4-t-Butylcyclohexyllithium (Li-00151), 145
N-t-Butyl-2,N-dilithiobenzamide, 84
4-t-Butyldimethylsilyloxy-2,6-dimethylphenyllithium, 145
2-(t-Butyldimethylsilyloxymethyl)-1-lithio-3,3-dimethyl-1-p-toluenesulphinyl-1-butene, 91
3-t-Butyl-3-hydroxy-2,2-dimethylheneicosane, 70
2-t-Butyl-2-lithio-1,3-dithiane, 91
N-t-Butyl-2-lithio-N-methylbenzenesulphonamide, 42, 81
4-t-Butyl-2-lithiothiophene, 111
n-Butyllithium (Li-30010)
  association, 3
  commercial availability, 21–22
  estimation, 17–18
  flammability, 11–12
  preparation, 25
  solubility, 5
s-Butyllithium (Li-00053)
  commercial availability, 21–22
  flammability, 11
t-Butyllithium (Li-00052)
  commercial availability, 22
  flammability, 11
  for metal–halogen exchange, 31
  preparation, 23, 48
3-Butyl-2-methylcyclohexanone, 73
4-t-Butyl-1-phenylthiocyclohexyllithium, 48–49
3-Butylpyridine, 108
2-t-Butylsulphonylphenyllithium, 136

C

Carbamates, 83, 86
Carbene complexes, 98–99
Carbenoids, 169ff
  β-oxido, 173
Carbodiimides, 57, 96
Carbonates, 79, 80
Carbon dioxide, 89ff
Carbon disulphide, 96, 101–103

INDEX OF COMPOUNDS AND METHODS

Carbon monoxide, 95ff
Carbonyls, 96, 98–99
Carboxylates, 89ff
Carboxylic acids, 89ff
Chiral ligands, 69
Chlorine, 143ff
α-Chlorobenzyllithium, 171
Chlorocarbamates, 83, 86
Chloro(2-chloroethoxy)methyllithium, 171
5-Chloro-2-(dimethylaminomethyl)phenyl-
  lithium, 41, 63
2-Chloro-3,3-diphenylacrylic acid, 92
1-Chloro-2,2-diphenylethenyllithium, 92
Chloroformates, 76, 77
3-Chlorofuran, 147
1-Chloro-2-iodoethane, 146
3-Chloro-3-lithiobut-1-ene, 171
2-Chloro-3-lithiopyridine, 168
3-Chloro-2-(methoxymethoxy)benzaldehyde, 87
3-Chloro-2-(methoxymethoxy)phenyllithium, 41, 87
α-Chloro-4-methylbenzyllithium, 171
3-(4-Chlorophenyl)-5-lithio-1-tetrahydro-
  pyran-2-ylimidazole, 68
2-Chlorophenyllithium (Li-10017), 168
3-Chloro-1-propynyllithium, 77
Chlorosilanes, 80, 96, 152–154
N-Chlorosuccinimide, 144
Comins' reagent, 83
Cooling baths, 8–9, 29
Cyanocuprates, 73, 96
Cyanocyclopentane, 64
4-Cyano-7,8-dimethoxy-1,2-dihydro-
  naphthalene, 63
Cyanogen bromide, 145
1-Cyclohexenyllithium (Li-00091), 25
5-(Cyclohexyliminomethyl)-3-ethoxy-2-
  methoxyphenyllithium, 150
2-(Cyclohexyliminomethyl)-4,5-dimethoxy-
  phenyllithium, 146
Cyclohexyllithium (Li-00094), 25
Cyclooctatetraenyllithium (Li-00119), 91
1-Cyclooctenyllithium, 132
Cyclopentanone enolate, 37
Cyclopentyllithium, 64
Cyclopropanes
  synthesis of, 169ff
Cyclopropyllithium (Li-00031), 25
Cycloreversion, 176, 178

D

DABCO, xvii, 4, 36, 38
Detection and estimation of organolithium
  compounds, 16ff
Deuteration, 39, 68, 120–123
Deuterium oxide, 39, 68, 120–123
1,4-Diazabicyclo[2.2.2]octane, see DABCO
1,2-Dibromoethane, 144
4,5-Dibromo-N-(ethoxy-
  methyl)-2-lithioimidazole, 30
Dibromofluoromethyllithium, 170
3,5-Dibromophenyllithium, 30
1,2-Dibromotetrachloroethane, 145
4,4'-Di-t-butylbiphenyl, 26
Di-t-butylketimine, 64
Dichloroallyllithium (Li-00028), 171
α,α-Dichlorobenzyllithium (Li-00103), 81, 84
1,1-Dichloroethyllithium, 80
2,2-Dichloro-4-methylpentan-3-one, 80
2,2-Dichloro-2-phenylethanal, 81
2,6-Dichlorophenyllithium, 168
2,6-Dicyanophenyllihtium, 145
Dicyclopentyl ketone, 64
3,3-Diethoxy-2-lithioprop-1-ene, 86
Diethyl ether, 5
N,N-Diethyl-2-lithiobenzamide, 39, 150
Diethyl α-lithioethylphosphonate, 138
N,N-Diethyl-(2-lithiophenyl)carbamate, 86
1,2-Diiodoethane, 146
N,N-Diisopropyl-2-lithiobenzamide
  (Li-30059), 122
1,2-Dilithiobenzene (Li-00086), 45
2,O-Dilithiobenzylalcohol, 138, 153
2,2'-Dilithiobiphenyl (Li-00156), 150, 161
Dilithio 4-chloroacetophenoneoxime, 78
3,3'-Dilithiodipropylether, 48
Dilithiohexadecanoic acid, 103, 138
1,3-Dilithio-1-hexyne, 40, 109
8,O-Dilithio-1-hydroxytetralin, 91
1,3-Dilithio-3-phenylbut-1-yne, 102
2,N-Dilithio-N-phenylethylamine, 46
1,4-Dilithio-1,2,3,4-tetraphenylbuta-1,3-
  diene (Li-00185), 161
1,3-Dilithio-8-thiabicyclo[4.3.0]nonane-
  S,S-dioxide, 109
2,5-Dilithiothiophene (Li-00041), 38
2,5-Dimethoxybenzylalcohol
  as indicator, 18

1,2-Dimethoxyethane, see DME
1,4-Dimethoxy-2-naphthyllithium, 99
2,6-Dimethoxyphenyllithium (Li-00121), 111, 138
3,6-Dimethoxy-2-(vinyloxymethyl) lithium, 146
Dimethylaminocarbonyllithium, 40
3-(Dimethylamino)-1-lithio-1-phenylthioprop-1-ene, 90
3-(Dimethylaminomethyl)-2-lithiothiophene, 145
Dimethylaminomethyllithium (Li-10008), 46
2-Dimethylamino-5-methylphenyllithium, 69
2-(Dimethylaminomethyl)phenyllithium (Li-00133), 68
4-(Dimethylaminomethyl)phenyllithium, 86
2-Dimethylaminophenyllithium (Li-00122), 41
3-(Dimethylamino)propyllithium (Li-30017), 25, 160
3,3-Dimethyl-1-butynyllithium (Li-30019), 136, 140, 144
2,3-Dimethylcyclopropyllithium, 114
N,N-Dimethylformamide, see DMF
4,6-Dimethylhept-1-en-4-ol, 70
Dimethyl(methylene)ammonium iodide, 58
2-(4,4-Dimethyloxazolin-2-yl)-3-lithiothiophene, 122
2-(4,4-Dimethyloxazolin-2-yl)-5-lithiothiophene, 122
2-(2,2-Dimethylpropyl)-2-lithio-1,3-dithiane, 54
2,2-Dimethylpropyllithium, see Neopentyllithium
1,6-Dimethyltricyclo[4.1.0.0^{2,7}]hept-3-ene, 172
Dinitrogen difluoride, 143
Diphenylacetic acid
  as indicator, 18
1,3-Diphenylacetone tosylhydrazone
  as indicator, 18
2-(Diphenylarsino)phenyllithium, 156
(2,2-Diphenylethenyl)diphenylphosphine sulphide, 75
1,2-Diphenylethenyllithium, 160
(Diphenylphosphino)methyllithium (Li-00167), 22, 156
2,4-Diphenylquinazoline, 59
(Diphenylthiophosphonyl)(trimethylsilyl)methyllithium, 75

Diselenides, 138
Disulphides, 96, 137–139
Disulphur dichloride, 139
1,3-Dithian-2-yllithium, see 2-Lithio-1,3-dithiane
5,9-Dithiaspiro[3.5]nonane, 110
Dithiocarbamates, 104
Dithiocarboxylates, 101–103
Dithioesters, 104
DME, xvii, 5, 7
DMF, xvii
  in synthesis of aldehydes, 82ff
  metallation of, 40

E

Enaminoketones, 86
Enolates
  generation of, 37, 129, 178
Epoxides
  reactions, 113ff
  synthesis, 169–171
5,8-Epoxy-1,2,3,4-tetrachloro-5,8-dihydronaphthalene, 168–169
Eschenmoser's salt, 58
Esters
  reactions, 76ff, 96
  synthesis, 76–77
2-Ethyl-1,3-dithiane, 112
Ethylene
  generation, 178
  metallation, 36
(E)-2-Ethyl-1-hexenyllithium, 113
2-Ethylhexyllithium, 48
Ethyl lithioacetate, 68
Ethyl α-lithiomalonate, lithium salt, 77
Ethyl 2-lithio-6-methoxybenzoate, 138
2-Ethyl-3-lithio-5-methylselenophene, 179
4-Ethyl-2-lithio-5-methylthiazole, 114
Ethyllithium (Li-00015)
  preparation, 24
  solubility, 5
(E)-4-Ethyloct-3-en-1-ol, 113
(Z)-2-(Ethylseleno)hept-2-en-4-yne, 179

F

N-Fluoro-N-t-butyl-p-toluenesulphonamide, 143

INDEX OF COMPOUNDS AND METHODS   185

2-Fluorophenyllithium (Li-00085), 168
N-Fluoroquinuclidinium fluoride, 143
Formamides, N,N-dialkyl, 82ff
Formate esters, 77ff
Formate, lithium, 93
N-Formylpiperidine, 83, 84
2-(2-Furyl)-2-lithio-1,3-dithiane, 178

G

Gilman colour tests, 16–17
Gilman double titration, 17–18
Grignard reagents, see Organomagnesium
  compounds

H

o-Halogenoaryllithium compounds, 167–169
1-Halogeno-1-lithioalkenes, 173
Hazards, 11ff, 143
Hexachloro-3,3′-dilithio-4,4′-bipyridyl, 31, 140
Hexachloroethane, 143, 145, 147
Hexametapol, xvii, 5, 6, 8, 36, 55, 72, 107, 108, 133
Hexamethylphosphoric triamide, see Hexametapol
Hexane, 5, 8
1-Hexenyllithium, 29, 137
1-Hexynyllithium, 153
Horner–Wadsworth–Emmons synthesis, 75, 157
Hydrazones, 57
Hydroperoxides, 119, 129ff
α-Hydroperoxyphenylacetic acid, 131
Hydroxylamine derivatives, 125ff
α-Hydroxyphenylacetic acid, 130
3-Hydroxy-2,2,3-trimethyloctan-4-one, 97

I

Imines
  addition of organolithium compounds, 57
  N-lithio, 62ff
  synthesis, 62ff
Iminium salts, 57–58

Indicators
  for titration of organolithium
    compounds, 18
Inert atmospheres, 9–10
Iodine, 143, 146, 147
Isocyanates, 89, 90, 96
Isonitriles, 65–66
2-(Isopropylthio)phenyllithium, 138, 156
Isoquinoline, 60
Isothiocyanates, 96, 103

K

Ketals, 178
Ketenes, 88, 90
Ketones
  reactions, 67ff, 96
  synthesis, 62ff, 76ff, 83ff, 89, 93

L

Lactones, 76ff
LDA (Li-20019), xvii, 35, 37, 108, 119, 167
LDMAN (Li-10037), xvii, 47–49
LiDBB (Li-00182), xvii, 26–27, 48
Lithiated O-(1-ethoxyethyl)-2-methyl-
  propanalcyanohydrin, 55
Lithiation, see Metallation
Lithioallene (Li-00029), 91
3-Lithiobenzo[b]furan (Li-00115), 177
2-Lithiobenzothiazole (Li-20021), 122
2-Lithiobenzo[b]thiophene (Li-30031), 90, 151
3-Lithiobenzo[b]thiophene (Li-00117), 31, 77
α-Lithiobenzylisonitrile, 117
2-(α-Lithiobenzyl)-4,4,6-trimethyldihydro-
  oxazine, 109
2-Lithio-3,3′-bithienyl, 127
2-Lithiobornene, 50
2-Lithio-1,3-butadiene (Li-20007), 50
6-Lithiochrysene, 120
3-Lithiocyclohex-1-ene, 48
4-Lithiodibenzofuran (Li-30051), 42, 130
3-Lithio-2,4-dimethoxyquinoline, 84
3-Lithio-2-(4,4-dimethyloxazolin-2-yl)thio-
  phene, 33
3-Lithio-2,5-dimethylthiophene, 177

2-Lithio-6,9-dioxaspiro[4.4]non-2-ene, 141
3-Lithio-2,5-diphenylfuran, 177
2-Lithio-4,5-diphenyloxazoline, 178
2-Lithio-1,3-dithiane (Li-10011)
  preparation, 36
  reactions, 63, 77, 109, 110, 114, 115
Lithio-1,4-dithiin, 177
2-Lithiofuran (Li-00042), 42, 63, 68, 77, 136
3-Lithiofuran (Li-00043), 29, 147
N-Lithioimines, 62
3-Lithio-2-(N-lithio-t-butylcarboxamido)-pyridine, 109
3-Lithio-2-(N-lithio-t-butylcarboxamido)-quinoline, 42
2-Lithio-N-(1-lithio-2,2-dimethyl-propylidene)aniline, 140
O-Lithio-2-(lithiomethyl)hexan-1-ol, 54
O-Lithio-2-(lithiomethyl)prop-1-en-2-ol, 40
1-Lithio-2-(2-lithiophenyl)-1-phenylhex-1-enyllithium, 153
1-Lithio-1-methoxyallene (Li-30009), 109, 114, 138
4-Lithio-5-methoxydibenzofuran, 136
4-Lithio-3-(methoxymethyl)pyridine, 122
α-Lithio-(4-methoxyphenyl)acetonitrile, 72
3-Lithio-2-methoxypyridine, 78
2-Lithio-5-methylbenzo[b]thiophene, 111
2-Lithio-2-methyl-1,3-dithiane (Li-10013), 63, 77, 84
3-Lithio-1-methylindazole, 177
Lithiomethylisonitrile, 102
(E)-1-Lithiomethyl-2-methoxycyclohexane, 48
1-Lithiomethyl-3-methylisoxazole, 102
2-Lithiomethyl-4-methylpyridine, 79
5-Lithio-1-methylpyrazole, 111
2-(Lithiomethyl)pyridine, see 2-Picolyllithium
3-(Lithiomethyl)pyridine (Li-00089), 108
2-Lithio-N-methylpyrrole (Li-00067), 42, 138
1-Lithio-1-methylthio-1,3-butadiene, 140
2-Lithio-N-nitrosopyrrolidine, 40, 68
7-Lithionorbornadiene (Li-20022), 27, 91, 122
(2-Lithio-3-pyrrolidinophenyl)-N,N-diethylcarbamate, 153
3-Lithio-2-pheylbenzo[b]thiophene, 28
3-Lithio-2-phenylchromone, 153

1-Lithio-2-phenyl-1,2-dihydropyridine, 60–61
2-(3-Lithiophenyl)-4,4-dimethyloxazoline, 121
2-Lithio-5-phenylfuran, 150
2-Lithio-5-phenyloxazole, 84
[Lithio(phenylthio)methylene]cyclohexane, 48, 153
2-Lithiothiophene (Li-00044), 38–39, 59, 68, 136
3-Lithiothiophene(Li-00045), 85, 132
3-Lithio-N-(triisopropylsilyl)pyrrole, 30
2-Lithio-1,3,5-trithiane, 109
Lithium
  atomic weight, 17
  reaction with ethers and thioethers, 47–48
  reactions with organic halides, 21ff
Lithium alkoxides, 119–120
Lithium amides, 119–120
  see also LDA, LiTMP
Lithium bis(trimethylsilyl)amide (Li-10020), 35, 120
Lithium 4,4'-di-t-butylbiphenyl, see LiDBB
Lithium t-butylperoxide, 119
Lithium n-butylsulphinate, 142
Lithium dibutylcuprate, 73
Lithium diisopropylamide, see LDA
Lithium 1-dimethylaminonaphthalene, see LDMAN
Lithium halides, 21–22, 36, 39, 67, 170
Lithium 2-lithioindole-1-carboxylate, 77, 78
Lithium lithiomethoxide, 46
Lithium 2-(lithiomethyl)allyloxide, 153
Lithium 2-(lithiomethyl)benzoate, 79
Lithium lithio(phenyl)acetate, 130
Lithium naphthalene (Li-00141), 26, 48
Lithium sulphinates, 141–142
Lithium tetraalkylborates, 149
Lithium 2,2,6,6-tetramethylpiperidide, see LiTMP
Lithium thiolates, 119–120, 135–136
LiTMP (Li-10028), xvii, 35, 120, 167

M

Magnesium bromide, 129, 163
  anhydrous, preparation, 163
Menthyllithium (Li-00152), 91

INDEX OF COMPOUNDS AND METHODS

Mesityllithium, see
2,4,6-Trimethylphenyllithium
Metal–halogen exchange, 27ff, 81, 99, 113, 123–124, 127, 155, 163, 170
  t-butyllithium as reagent for, 29
  in presence of reactive functional groups, 27–28
  in presence of tritium oxide, 121
Metallation, 32ff, 80, 92, 154
  of heterocycles, 36, 38, 85, 151
  ortho, 33ff, 39, 86, 87
  reagents for, 35
Methoxyamine, 124–125
5-Methoxy-2-(4,4-dimethyloxazolin-2-yl)phenyllithium, 41
Methoxymethyllithium (Li-00016), 153
2-(Methoxymethoxy)phenyllithium, 145
2-Methoxyphenyllithium (Li-00019), 78, 91
3-Methoxyphenyllithium, 132
2-(4-Methoxyphenyl)-2-(3-oxocyclohexyl)-acetonitrile, 72
1-Methoxyvinyllithium (Li-00034), 40,46
(η^6-2-Methylanisole)tricarbonylchromium, 55
1-d-2-Methylbutanal, 121–123
Methyl 2-carboxydithiohexadecanoate, 103
Methyldiphenylbismuth, 163
2-Methylene-1,3-dithiane, 54
N-Methylformanilide, 83
2-Methylhexan-1-ol, 54
Methyllithium (Li-30003)
  association, 3
  commercial availability, 22
  isotopically labelled, 30, 161
  preparation, 23, 25, 30
  solubility, 5
2-Methyl-5-(2-methylpropanoyl)anisole, 55
4-Methyl-4-phenylpentan-2-one, 73
[Methyl(phenyl)thiophosphonyl]methyl-lithium, 78
N-Methyl-N-(2-pyridyl)formamide, 83
2-Methylseleno-2-propyllithium, 46
α-(Methylthio)benzyllithium, 40, 109, 156
Methylthiomethyllithium (Li-10006), 156
[2-(Methylthio)phenyl]diphenylphosphine, 155
2-(Methylthio)phenyllithium (Li-30026), 30, 155, 156
Molybdenum pentoxide-pyridine-hexametapol (MoOPH), xvii, 133

N

Neopentyllithium (Li-30014), 22, 25, 162
Nitrates, 125
Nitriles, 62–65
Nitrogen, 9

O

(E)-Oct-2-ene, 175
Organoaluminium compounds, 160
Organoarsenic compounds, 161
Organobismuth compounds, 163
Organoboron compounds, 149ff
Organocadmium compounds, 163
Organocerium compounds, 69, 163
Organochromium compounds, 162
Organocopper compounds, 54, 57, 73–74, 80, 96, 108, 161, 163
Organogermanium compounds, 160
Organogold compounds, 161
Organohafnium compounds, 69, 163
Organoiridium compounds, 162
Organolithium compounds
  association, 3–4
  commercial suppliers, 21–22
  detection and estimation, 16–18
  hazards, 11–12
  large-scale suppliers, 21
  reactions, see under individual substrate types
  solvents for, 5ff, 22
  storage and handling, 11–12
Organomagnesium compounds, 129–130, 163
Organomanganese compounds, 69, 80, 163
Organomercury compounds, 45–46, 160
Organopalladium compounds, 108, 163
Organophosphorus compounds, 154ff
Organopotassium compounds, 36, 39, 163
Organoselenium compounds, 45, 46, 135ff
Organosilicon compounds, 74, 152–154
Organosilver compounds, 161
Organosodium compounds, 163
Organotantalum compounds, 162
Organothallium compounds, 160
Organotin compounds, 44, 46, 161
Organotitanium compounds, 69, 162, 163
Organovanadium compounds, 162

INDEX OF COMPOUNDS AND METHODS

Organozinc compounds, 160, 163
Orthoesters, 107
Oxetanes, 116–117
Oxiranes, *see* Epoxides
Oxygen, 129–131

P

Pentacarbonyl[methoxy(1,4-dimethoxy-2-naphthyl)carbene]chromium(o), 99
Pentachloroiodobenzene, 147
Pentachlorophenyllithium (Li-00080), 29, 63, 91, 114, 117, 147, 168–169
3-Pentachlorophenylpropan-1-ol, 117
Pentafluorophenyllithium (Li-00081), 30, 140, 150, 161
Pentane, 5, 8
Perchloryl fluoride, 143
Peroxides, 131–132
Peterson olefination, 74–75
Phenanthridine, 59
1,10-Phenanthroline as indicator, 18
Phenylethynyllithium (Li-00113), 150
Phenyllithium (Li-10018)
  association, 3
  commercial availability, 22
  preparation, 24, 46
N-Phenyl-1-naphthylamine as indicator, 18
3-Phenylpyridazine, 59
2-Phenylpyridine, 60
Phenylselenomethyllithium (Li-10023), 46
(E)-1-Phenylthio-1-hexene, 137–138
Phenylthiomethyllithium (Li-10022), 153
1-Phenylthio-1-trimethylsilylethene, 154
1-Phenylthiovinyllithium, 40, 154
2-Phenyl-1-(trimethylsilyl)ethyllithium, 140
Phosphates, 112, 157
Phosphine oxides, 75
Phosphine sulphides, 75
Phosphinites, 155
Phosphites, 155, 156
Phosphorus halides, 155ff
2-Picolyllithium (Li-10019), 63, 68
Potassium t-butoxide, 36,39
Potassium diisopropylamide, 36
1-Propynyllithium (Li-00030), 136
Pyrazine, 59
Pyridine, 7, 58–60
2-Pyridyllithium (Li-0060), 31
Pyrimidine, 59

Q

Quinoline, 59

R

Radical anions, 26,47–49

S

Schlenk tubes, 11, 14
Selenenyl halides, 140
Silyl halides, 152–154
Shapiro reaction, 50–51
Sodium–lithium alloys, 23
Solvents, 22
  drying of, 5–7
  effect on constitution of organolithium compounds, 3–4
  effect on reactivity of organolithium compounds, 3–4, 35–36, 67
Sulphates, 28, 111, 112
Sulphinates, 141
Sulphinyl halides, 141
Sulphonates, 112, 141
Sulphones
  $\alpha\beta$-unsaturated, 53
Sulphonyl halides, 141, 142, 144, 145
Sulphonylhydrazones, 49–50
Sulphur, 96, 135–137
Sulphur dichloride, 139–140
Sulphur dioxide, 141–142
Sulphuryl chloride, 141

T

1,1,2,2-Tetrabromoethane, 145
Tetrachloro-3-lithiopyridine, 144
Tetrahydrofuran, *see* THF
2,3,5,6-Tetramethoxyphenyllithium, 84
1,1,3,3-Tetramethylbutylisonitrile, 65–66
N,N,N',N'-Tetramethyl-1,2-ethanediamine, *see* TMEDA
THF, xvii, 5–8
  cleavage of, 35, 116, 176, 178
  drying of, 5–7
Thioamides, 103

INDEX OF COMPOUNDS AND METHODS

Thiocyanates, 139–140
Thioethers
  cleavage of, 47–49
  synthesis, 135
Thioketenes, 105
Thioketones, 104–105
Thiophene-2-thiol, 136–137
Thiophilic addition, 104–105
TMEDA, xvii, 3, 4, 7, 22, 36, 39, 54, 80, 87, 154
2-(p-Toluenesulphonyl)-2-propyllithium, 102
4-Tolyllithium (Li-00108), 22, 94
Transmetallation, 44ff
Trapp mixture, 8, 137
Triazenes, 127
2,4,6-Tri-t-butylphenyllithium (Li-20054) 157
4,4,4-Triethoxy-1-butynyllithium, 78
Triethylamine, 5, 7
2,4,6-Trimethylphenyllithium (Li-00132), 30, 162
1-(Trimethylsilyl)ethyllithium, 140
1-(Trimethylsilyl)-1-hexenyllithium, 30
Trimethylsilylmethyllithium (Si-30007), 22, 25, 161
2-(Trimethylsilyloxy)vinyllithium, 30

4-(Trimethylsilyl)phenyllithium, 30, 164
Triphenylmethane as indicator, 18
Triphenylmethyllithium (Li-30069), 40
Trithiocarbonates, 104
Tritiation, 28, 120–121

U

Ultrasonic irradiation, 23, 83

V

Vinyllithium (Li-30006), 25, 46, 94

W

Wittig reactions, 174–175

Y

Ylides, 174–175